Lecture Notes
in Computational Science
and Engineering

133

Editors:

Timothy J. Barth
Michael Griebel
David E. Keyes
Risto M. Nieminen
Dirk Roose
Tamar Schlick

More information about this series at http://www.springer.com/series/3527

Harald van Brummelen • Cornelis Vuik •
Matthias Möller • Clemens Verhoosel •
Bernd Simeon • Bert Jüttler
Editors

Isogeometric Analysis and Applications 2018

 Springer

Editors
Harald van Brummelen
Department of Mechanical Engineering
Eindhoven University of Technology
Eindhoven, The Netherlands

Cornelis Vuik
Department of Applied Mathematics
Delft University of Technology
Delft, The Netherlands

Matthias Möller
Department of Applied Mathematics
Delft University of Technology
Delft, The Netherlands

Clemens Verhoosel
Department of Mechanical Engineering
Eindhoven University of Technology
Eindhoven, The Netherlands

Bernd Simeon
Fachbereich Mathematik
Technische Universität Kaiserslautern
Kaiserslautern, Germany

Bert Jüttler
Institut für Angewandte Geometrie
Johannes Kepler Universität Linz
Linz, Austria

ISSN 1439-7358   ISSN 2197-7100 (eBook)
Lecture Notes in Computational Science and Engineering
ISBN 978-3-030-49835-1   ISBN 978-3-030-49836-8 (eBook)
https://doi.org/10.1007/978-3-030-49836-8

Mathematics Subject Classification: 65D05, 65D07, 65D17, 65D32, 65M38, 65M50, 65M55, 65M60, 65N30, 65N38, 65N55

© Springer Nature Switzerland AG 2021

This work is subject to copyright. All rights are reserved by the Publisher, whether the whole or part of the material is concerned, specifically the rights of translation, reprinting, reuse of illustrations, recitation, broadcasting, reproduction on microfilms or in any other physical way, and transmission or information storage and retrieval, electronic adaptation, computer software, or by similar or dissimilar methodology now known or hereafter developed.

The use of general descriptive names, registered names, trademarks, service marks, etc. in this publication does not imply, even in the absence of a specific statement, that such names are exempt from the relevant protective laws and regulations and therefore free for general use.

The publisher, the authors, and the editors are safe to assume that the advice and information in this book are believed to be true and accurate at the date of publication. Neither the publisher nor the authors or the editors give a warranty, expressed or implied, with respect to the material contained herein or for any errors or omissions that may have been made. The publisher remains neutral with regard to jurisdictional claims in published maps and institutional affiliations.

This Springer imprint is published by the registered company Springer Nature Switzerland AG.
The registered company address is: Gewerbestrasse 11, 6330 Cham, Switzerland

Participants of the 3rd Conference on Isogeometric Analysis and Applications (IGAA 2018) held April 23rd–26th, 2018 in Delft, The Netherlands

Foreword

This book contains a selection of papers emanating from the third workshop in the **IGAA** series, **Isogeometric Analysis and Applications—IGAA 2018**, held at the Science Centre Delft, on the campus of the Delft University of Technology (TU Delft), Monday, April 23rd through Thursday, April 26th, 2018 in the historic and picturesque city of Delft, home of the father of microbiology, Antonie van Leeuwenhoek, 1632–1723, the peerless painter Johannes Vermeer, 1632–1675, and the beautiful blue and white Delft ceramic pottery. The program consisted of five plenary lectures and 45 contributed presentations and a poster session and reception at the conclusion of the first day's presentations. The third day of the conference was designated Industry Day and concluded with a lively panel session entitled The Future of IGA in Industry. This was the first session of its kind at an IGA conference but certainly will not be the last. We expect each subsequent conference devoted to Isogeometric Analysis to adopt similar themes for sessions and panel discussions due to the increased interest in the use of IGA in industry and concomitantly among the providers of commercial software tools. The social highlight of the conference was the excellent banquet held at the Michelin-starred restaurant Aan de Zweth.

The plenary lectures were given by Josef Kiendl of the Norwegian University of Science and Technology, Kenji Takada of the Honda Motor Company, Jessica Zhang of Carnegie Mellon University, Carlotta Giannelli of the University of Florence, and the writer. After opening remarks, the technical sessions began with my plenary lecture in which I presented an overview of the analytical attributes of IGA vis-á-vis classical finite element analysis (FEA) and some illustrations of the applicability to problems whose features are not amenable to solution by classical FEA. Josef Kiendl presented a summary of his research in thin shell structural analysis, an area that he pioneered and one in which IGA has created a renaissance. Kenji Takada described the uses of IGA at Honda Motor Company and showed the efficiency and accuracy benefits of IGA in real-world automotive engineering applications. Kenji amused the audience with the story of how he serendipitously began his study of IGA. Jessica Zhang described her extensive work on isogeometric model development and many recent industrial applications. Carlotta Giannelli presented her mathematical research in adaptive IGA schemes utilizing hierarchical splines, an important technology for the efficient solution of practical problems.

The contributed presentations likewise spanned the very mathematical through the very applied, and everything in between, a hallmark of IGA. It seems there is something for everyone. However, what occurred at **IGAA 2018** was a much greater focus on the industrial sector. This definitely appears to be a new thrust in IGA and we can look forward toward much greater interaction with industry in the future. It seems too that many industrial users are calling for the development of IGA modeling tools so that they can use IGA solution capabilities that now exist in commercially available FEA codes, such as LS-DYNA, and several companies in the mesh and model generation area have begun to respond to this opportunity.

At the same time, IGA research is still a growing and vibrant area. The subject is rich with possibilities of new developments, even breakthroughs. I am personally always amazed when classical areas of analysis are addressed anew by IGA and fundamentally new technologies emerge that offer unique combinations of accuracy, ease of use, and efficiency. IGA also has the potential to open entirely new application areas, especially ones where smooth basis functions are desirable or even necessary. The unique connections embodied in the IGA theme between design and analysis provide a strong platform for future widespread adoption in industry and commercialization.

The papers in this volume could only sample some of the broad spectrum of topics presented at the meeting. Nevertheless, they give an indication of the breadth of the subject and provide important technical advancements in several areas. Like its predecessors in the **IGAA** workshop series, this proceedings volume is an important contribution to the growing scientific literature on Isogeometric Analysis.

Many thanks are due to the organizers of the workshop for their very hard work which produced a meeting of the highest scientific quality and one that was truly a great pleasure to attend.

Austin, TX, USA
June 2019

Thomas J. R. Hughes

Preface

With 66 participants from 14 countries, the **3rd Conference on Isogeometric Analysis and Applications 2018** was a successful continuation of the IGAA conference series that was started in 2012 in Linz, Austria and had its second edition in 2014 in Annweiler am Trifels, Germany. IGAA 2018 has brought together senior experts from academia and industry as well as young scientists standing at the beginning of their career to discuss the current state of the art in IGA and its potential for commercial application in industry. This book contains a selection of 12 papers addressing various aspects of IGA ranging from the mathematical analysis of approximation properties over the generation of analysis-suitable parametrizations and the efficient solution of linear systems of equations, to its use in practical applications.

A conference like IGAA cannot be organized without external support. The organizers would like to thank the 4TU Applied Mathematics Institute (4TU.AMI), the TU Delft Institute for Computational Science and Engineering (DCSE), the J.M. Burgerscentrum, and the NWO Cluster Nonlinear Dynamics of Natural Systems (NDNS+) for generous financial support. Financial support from the European Commission through project MOTOR (GA no. 678727) is also greatly acknowledged. We finally would like to express our gratitude to Mrs. Deborah Dongor and her colleagues for taking care of the local organization of this event.

Eindhoven, The Netherlands	Harald van Brummelen
Linz, Austria	Bert Jüttler
Delft, The Netherlands	Matthias Möller
Kaiserslautern, Germany	Bernd Simeon
Eindhoven, The Netherlands	Clemens Verhoosel
Delft, The Netherlands	Cornelis Vuik
January 2020	

Contents

Generating Star-Shaped Blocks for Scaled Boundary Multipatch IGA ... 1
Benjamin Bauer, Clarissa Arioli, and Bernd Simeon

Approximation Power of C^1-Smooth Isogeometric Splines on Volumetric Two-Patch Domains ... 27
Katharina Birner, Bert Jüttler, and Angelos Mantzaflaris

A Novel Approach to Fluid-Structure Interaction Simulations Involving Large Translation and Contact ... 39
Daniel Hilger, Norbert Hosters, Fabian Key, Stefanie Elgeti, and Marek Behr

An IGA Framework for PDE-Based Planar Parameterization on Convex Multipatch Domains ... 57
Jochen Hinz, Matthias Möller, and Cornelis Vuik

Preconditioning for Linear Systems Arising from IgA Discretized Incompressible Navier–Stokes Equations ... 77
Hana Horníková and Cornelis Vuik

Solving 2D Heat Transfer Problems with the Aid of a BEM-Isogeometric Solver ... 99
Konstantinos Kostas, Yeraly Kalel, and Azat Amiralin

Isogeometric Methods for Free Boundary Problems ... 131
M. Montardini, F. Remonato, and G. Sangalli

Approximately C^1-Smooth Isogeometric Functions on Two-Patch Domains ... 157
Agnes Seiler and Bert Jüttler

Properties of Spline Spaces Over Structured Hierarchical Box Partitions ... 177
Ivar Stangeby and Tor Dokken

Efficient p-Multigrid Based Solvers for Isogeometric Analysis on Multipatch Geometries 209
Roel Tielen, Matthias Möller, and Cornelis Vuik

The Use of Dual B-Spline Representations for the Double de Rham Complex of Discrete Differential Forms 227
Yi Zhang, Varun Jain, Artur Palha, and Marc Gerritsma

Manifold-Based B-Splines on Unstructured Meshes 243
Qiaoling Zhang, Thomas Takacs, and Fehmi Cirak

Contributors

Azat Amiralin School of Engineering, Nazarbayev University, Astana, Kazakhstan

Clarissa Arioli TU Kaiserslautern, Dept. of Mathematics, Felix Klein Zentrum für Mathematik, Kaiserslautern, Germany

Benjamin Bauer Fraunhofer ITWM, Kaiserslautern, Germany

Marek Behr Chair for Computational Analysis of Technical Systems, RWTH Aachen University, Aachen, Germany

Katharina Birner Institute of Applied Geometry, Johannes Kepler University Linz, Linz, Austria

Fehmi Cirak Department of Engineering, University of Cambridge, Cambridge, UK

Tor Dokken SINTEF, Oslo, Norway

Stefanie Elgeti Chair for Computational Analysis of Technical Systems, RWTH Aachen University, Aachen, Germany

Marc Gerritsma Faculty of Aerospace Engineering, Delft University of Technology, Delft, The Netherlands

Daniel Hilger Chair for Computational Analysis of Technical Systems, RWTH Aachen University, Aachen, Germany

Jochen Hinz Delft Institute of Applied Mathematics, Delft University of Technology, Delft, The Netherlands

Hanah Horníková Faculty of Applied Sciences, University of West Bohemia, Plzeň, Czech Republic

Norbert Hosters Chair for Computational Analysis of Technical Systems, RWTH Aachen University, Aachen, Germany

Thomas J. R. Hughes Institute for Computational Engineering and Sciences (ICES), The University of Texas at Austin, Austin, TX, USA

Varun Jain Faculty of Aerospace Engineering, Delft University of Technology, Delft, The Netherlands

Bert Jüttler Institute of Applied Geometry, Johannes Kepler University Linz, Linz, Austria

Yeraly Kalel School of Engineering, Nazarbayev University, Astana, Kazakhstan

Fabian Key Chair for Computational Analysis of Technical Systems, RWTH Aachen University, Aachen, Germany

Konstantinos Kostas School of Engineering, Nazarbayev University, Astana, Kazakhstan

Angelos Mantzaflaris Institute of Applied Geometry, Johannes Kepler University Linz, Linz, Austria
Research Centre Inria Sophia Antipolis - Méditerranée, Université Côte d'Azur, Nice, France

Matthias Möller Delft Institute of Applied Mathematics, Delft University of Technology, Delft, The Netherlands

Monica Montardini Department of Mathematics, University of Pavia, Pavia, Italy

Artur Palha Faculty of Aerospace Engineering, Delft University of Technology, Delft, The Netherlands

Filippo Remonato Department of Mathematics, University of Pavia, Pavia, Italy
Department of Mathematical Sciences, NTNU, Trondheim, Norway

Giancarlo Sangalli Department of Mathematics, University of Pavia, Pavia, Italy
IMATI-CNR "E. Magenes", Pavia, Italy

Agnes Seiler Doctoral Program Computational Mathematics, Johannes Kepler University Linz, Linz, Austria

Bernd Simeon TU Kaiserslautern, Department of Mathematics, Felix Klein Zentrum für Mathematik, Kaiserslautern, Germany

Ivar Stangeby SINTEF, Oslo, Norway

Thomas Takacs Institute of Applied Geometry, Johannes Kepler University Linz, Linz, Austria

Roel Tielen Delft Institute of Applied Mathematics, Delft University of Technology, Delft, The Netherlands

Cornelis Vuik Delft Institute of Applied Mathematics, Delft University of Technology, Delft, The Netherlands

Qiaoling Zhang Department of Engineering, University of Cambridge, Cambridge, UK

Yi Zhang Faculty of Aerospace Engineering, Delft University of Technology, Delft, The Netherlands

Generating Star-Shaped Blocks for Scaled Boundary Multipatch IGA

Benjamin Bauer, Clarissa Arioli, and Bernd Simeon

Abstract This paper deals with the decomposition of a domain into star-shaped blocks, which is motivated by the idea of solving PDEs by means of the scaled boundary isogeometric analysis (SB-IGA). In the first part of the paper an introduction to the SB-IGA is given and we show the necessity for the domain to be star-shaped. Of course not every domain has this property, and for this reason the main focus of the paper is the generation of star-shaped blocks. The approaches that we take into account are the quadtree decomposition and the art gallery decomposition. We highlight the steps of those algorithms and we provide a computable sufficient criterion for the identification of the star-shapedness of a block. Moreover, for the art gallery decomposition an extension of Fisk's method and the Voronoi diagram are employed. Finally we solve the Poisson equation on different geometries, among them the Yeti footprint domain, and compare the approaches.

1 Introduction

Despite the significant progress in Isogeometric Analysis (IGA) over the last decade, the development of analysis-suitable parametrizations of the computational domain is still a major issue. In this paper, we investigate a special class of such parametrizations that are based on the decomposition of a general two-dimensional domain into star-shaped blocks in combination with the scaled boundary approach [1]. The latter can be viewed as a generalization of classical polar coordinates and is

B. Bauer (✉)
Fraunhofer ITWM, Kaiserslautern, Germany
e-mail: benjamin.bauer@itwm.fraunhofer.de

C. Arioli · B. Simeon
TU Kaiserslautern, Dept. of Mathematics, Kaiserslautern, Germany
e-mail: clarissa.arioli@gmail.com; simeon@mathematik.uni-kl.de

© Springer Nature Switzerland AG 2021
H. van Brummelen et al. (eds.), *Isogeometric Analysis and Applications 2018*,
Lecture Notes in Computational Science and Engineering 133,
https://doi.org/10.1007/978-3-030-49836-8_1

easy to construct as long as the domain is star-shaped. If not, a multipatch approach is required that decomposes the domain into appropriate blocks.

For this purpose, we make use of two alternative ideas: the quadtree decomposition and methods that stem from the classical art gallery problem. The quadtree decomposition is a standard in computer science and provides a fast search for neighbouring quadtree cells, which in our context means that each interface of the resulting multipatch structure lies on an edge between two cells in the tree. Furthermore, the concept is easily generalizable to higher dimensions, see [6, Chap. 14] for a comprehensive exposition. On the other hand, the art gallery problem [21] leads to a different approach where typically only polygonal boundaries are considered. In both cases, it is essential to guarantee the star-shapedness of the resulting blocks in a computationally feasible way, and we provide criteria and corresponding tools that can be implemented in the standard framework of splines, i.e., using the data given by the control polygons and knot vectors. The computational examples demonstrate that both approaches can be employed in a rather automatic fashion. Nevertheless, there are some pitfalls and worst cases that we also discuss.

In order to give a short overview on the state-of-the-art in the field, we refer to the original work by Hughes et al. [11, 5] that set the ball rolling in IGA and to Xu et al. [27] and Gravesen et al. [9] for results on the design and analysis of parametrizations in IGA. The Scaled-Boundary IGA (SB-IGA) is inspired by the Scaled Boundary Finite Element Method (SB-FEM) [3, 24, 25] and has been introduced in [2, 4, 15, 20]. The SB-FEM has been recently combined with quadtree-type decompositions in [8, 16]. Multipatch parametrizations and their treatment in IGA are addressed, e.g., in [10, 13, 18]. Last but not least, the visibility of continuous curves, which is related to the concept of star-shaped, has recently been studied in [12].

The paper is organized as follows. In Sect. 2, we summarize the scaled boundary approach. Section 3 concentrates on the quadtree algorithm and its application to our framework. This includes results that show how star-shapedness with respect to the control polygon and with respect to the boundary curve are related to each other. Moreover, a fast algorithm for the intersection of the boundary curve with a vertical or horizontal ray is given. The methods for the art gallery problem are adopted in Sect. 4. We concentrate on a generalized version of Fisk's algorithm, illustrate the corresponding framework of optimization in graphs and transfer the polygon tessellation to the domain by an approach related to Voronoi diagrams. Finally, Sect. 5 presents computational examples, among them the well-known Yeti footprint [14].

2 Scaled Boundary Isogeometric Analysis

In Computed-Aided Geometric Design (CAGD), objects are typically modelled in terms of their inner and outer hull, i.e., only a surface description is generated. On the other hand, for the purpose of applying IGA, one needs a computational mesh

2.1 Tensor Product B-Splines

Given a polynomial degree q, we introduce the knot vector

$$\Xi = \{\xi_1 \leq \xi_2 \leq \ldots \leq \xi_{n+q+1}\},$$

which contains non-decreasing parametric real values so that $0 \leq \mu(\Xi, \xi) \leq q + 1$ is the multiplicity of the parameter value in the knot vector (the multiplicity $\mu(X, x)$ is zero if the given value x is not a knot in X). Denoting the univariate B-splines of degree q by $\mathcal{N}_{j,q}(\xi)$, $j = 1, \ldots, n$ (we refer to [7] for their construction), we define a B-spline curve $\vartheta : [\xi_1, \xi_{n+q+1}] \to \mathbb{R}^2$ via

$$\vartheta(\xi) := \sum_{j=1}^{n} \mathcal{N}_{j,q}(\xi) P_j,$$

where $P_j \in \mathbb{R}^2$ are the control points. The polygonal chain constructed with the control points is called the control polygon $\mathcal{P}$. We recall now two important properties of B-spline curves:

- Strong convex hull property: the B-spline curve is contained in the convex hull of its control polygon.
- Variation diminishing property: each hyperplane in $\mathbb{R}^n$ has at most as many intersections with the B-spline curve as with the control polygon.

Furthermore, our algorithms will take advantage of the very easy splitting and merging process of B-spline curves. By knot insertion (cf. [22, Chap. 5]) the multiplicity of a knot $\hat{\xi}$ can be incremented until $\mu(\hat{\xi}) = q + 1$. Then there occurs a discontinuity in $\hat{\xi}$ in which the curve is split. Merging works the same way backwards.

Considering another univariate B-splines $\mathcal{M}_{i,p}(\eta)$, $i = 1, \ldots, m$, with knot vector

$$\Psi = \{\eta_1 \leq \eta_2 \leq \ldots \leq \eta_{m+p+1}\},$$

and multiplicity

$$0 \leq \mu(\Psi, \eta) \leq p + 1,$$

a function $f(\xi, \eta) : [\xi_1, \xi_{n+q+1}] \times [\eta_1, \eta_{m+p+1}] \to \mathbb{R}^2$ is called a bivariate tensor product B-spline function if it has the form

$$f(\xi, \eta) = \sum_{i=1}^{m} \sum_{j=1}^{n} \mathcal{M}_{i,p}(\eta) \mathcal{N}_{j,q}(\xi) Q_{i,j} \qquad (1)$$

with control points $Q_{i,j} \in \mathbb{R}^2$ forming the control mesh (or control net).

2.2 Scaled Boundary (SB) Parametrization

Consider a domain $\Omega \subset \mathbb{R}^2$ and its parametrization by a global geometry function

$$F : \Omega_0 \to \Omega, \quad F(\xi, \eta) = x \in \mathbb{R}^2, \qquad (2)$$

see Fig. 1. F is an invertible C^1-mapping from the parameter domain $\Omega_0 \subset \mathbb{R}^2$ to the physical domain Ω and, in our framework, $\Omega_0 = [0, 1]^2$ is the unit square.

Integrals over Ω can be transformed into integrals over Ω_0 via the integration rule

$$\int_{\Omega} g(x) \, dx = \int_{\Omega_0} g(F(\xi, \eta)) \, |\det DF(\xi, \eta)| \, d(\xi, \eta)$$

with 2×2 Jacobian matrix $DF(\xi, \eta)$. Next, we assume that Ω is a star-shaped domain whose boundary is described by a spline curve

$$\gamma(\eta) = \sum_{i=1}^{m} \mathcal{M}_{i,p}(\eta) P_i \qquad (3)$$

with univariate B-splines $\mathcal{M}_{i,p}$ of a certain degree p and control points $P_i \in \mathbb{R}^2$. We require $\eta \in [0, 1]$ and consider a closed curve with $\gamma(0) = \gamma(1)$. This can be achieved by an open knot vector and control points $P_1 = P_m$.

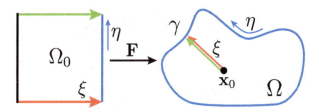

Fig. 1 Basic idea of an SB-parametrization

Choosing a point $x_0 \in \Omega$, the *scaling center*, the geometry function F for a scaled boundary parametrization (SB-parametrization) is defined as follows

$$F(\xi, \eta) = x_0 + \xi(C \cdot M(\eta) - x_0), \qquad (4)$$

where the matrix $C := (P_1, \ldots, P_m) \in \mathbb{R}^{2 \times m}$ contains the control points on the boundary and for all η the vector $M(\eta) := (\mathcal{M}_{1,p}(\eta), \ldots, \mathcal{M}_{m,p}(\eta))^T \in \mathbb{R}^m$ the B-splines.

Another equivalent representation to (4) is given by the tensor product structure (1) with only $n = 2$ linear B-splines in ξ-direction and m B-splines inherited from the boundary curve in η-direction. Therefore we have

$$F(\xi, \eta) = \sum_{i=1}^{m} \sum_{j=1}^{2} \mathcal{M}_{i,p}(\eta) \mathcal{N}_{j,1}(\xi) Q_{i,j},$$

with linear B-splines $\mathcal{N}_{1,1}(\xi) = 1 - \xi$, $\mathcal{N}_{2,1}(\xi) = \xi$ and control points Q_{ij} given by $Q_{i,1} := x_0$, $Q_{i,2} := P_i$, $i = 1, \ldots, m$. Finally, it is possible to refine the mesh structure in radial direction ξ by applying knot insertion and degree elevation, which leads to a representation of F in the general bivariate form (1).

2.3 SB-Parametrizations in Isogeometric Analysis

We now outline the use of SB-parametrizations in IGA. For simplicity, we consider Poisson's equation

$$-\Delta u = f \quad \text{in } \Omega \qquad (5)$$

as a model problem. Here, $\Omega \subset \mathbb{R}^2$ is a domain with boundary $\partial \Omega$, the function $f : \Omega \to \mathbb{R}$ is a given source term, and the unknown function $u : \Omega \to \mathbb{R}$ shall satisfy the Dirichlet boundary condition

$$u = 0 \quad \text{on } \partial \Omega. \qquad (6)$$

The weak form of the PDE (5) is obtained by multiplication with test functions w and integration over Ω. More precisely, one defines the function space $W := \{w \in H^1(\Omega), w = 0 \text{ on } \partial \Omega\}$, which consists of all functions $w \in L_2(\Omega)$ that possess weak and square-integrable first derivatives and that vanish on the boundary. For functions $u, w \in W$, the bilinear form

$$a(u, w) := \int_{\Omega} \nabla u \cdot \nabla w \, dx$$

is well-defined, and even more, it is symmetric and coercive. Setting

$$\langle l, w \rangle := \int_\Omega fw \, dx$$

as linear form for the integration of the right hand side, the solution $u \in W \subset H^1(\Omega)$ is then characterized by the weak form

$$a(u, w) = \langle l, w \rangle \quad \text{for all } w \in W \tag{7}$$

and the boundary condition $u = 0$ (in the sense of traces).

Next assume that we have a parametrization of Ω available as in (2). For the differentiation, the chain rule applied to $u(x) = u(F(\xi, \eta)) =: \hat{u}(\xi, \eta)$ yields, using a row vector notation for the gradient ∇u,

$$\nabla_x u(x) = \nabla_{\xi,\eta} \hat{u}(\xi, \eta) \cdot DF(\xi, \eta)^{-1}.$$

The integrals in the weak form (7) are then transformed to parametric coordinates, which yields

$$\int_{\Omega_0} (\nabla \hat{u} \, DF^{-1}) \cdot (\nabla \hat{w} \, DF^{-1}) \, |\det DF| \, d(\xi, \eta) = \int_{\Omega_0} \hat{f} \hat{w} \, |\det DF| \, d(\xi, \eta).$$

Note that for the discretization via the usual Galerkin projection, the two-dimensional integration on the left-hand side can be carried out as the product of two one-dimensional integrations, which is a great computational advantage of the scaled boundary parametrization when the stiffness matrix is assembled, see [1] for the details.

2.4 Multipatch Geometries

For the usage of the scaled boundary IGA one needs a star-shaped domain and, clearly, not every domain has this property, which we define properly in Sect. 3.1. For this reason, we decompose first the initial domain into subdomains or patches such that each of them is star-shaped. The notation for such a decomposition is highlighted in Fig. 2. We denote by $\Omega^{(k)}$ the k-th sub-domain and by $\Gamma^{(k,l)}$ the boundary or interface between $\Omega^{(k)}$ and $\Omega^{(l)}$.

The problem (5)–(6) can then be rewritten in the following way:

$$-\Delta u^{(k)} = f^{(k)} \quad \text{in } \Omega^{(k)} \quad \forall k, \tag{8a}$$

$$u^{(k)} = 0 \quad \text{on } \partial \Omega^{(k)} \quad \forall k, \tag{8b}$$

$$u^{(k)} = u^{(l)} \quad \text{on } \Gamma^{(k,l)} \quad \forall k, l. \tag{8c}$$

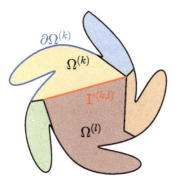

Fig. 2 Multipatch domain decomposition

The conditions (8c) are required to obtain $\mathscr{C}^0$ continuity between each pair of connected sub-domains. To achieve higher global continuity of the solution, one needs further conditions on the gradients along the interface, but the focus of our work is not an enhancement of this global continuity. Hence, in our examples in Sect. 5 we concentrate on $\mathscr{C}^0$ continuity. For $\mathscr{G}^1$ continuity the reader is referred to, e.g., [13], and for multipatch discontinuous Galerkin IGA see [18].

3 Quadtree Decomposition

Basically, a quadtree is a tree data-structure, where every node has either exactly four children or none. The latter ones are called leaves. It can be utilized to store planar data, for example images or point sets, or for mesh generation. In this process, each node of the tree represents a square region, where child nodes correspond to the four quadrants of the square of their parents. These quadrants are often referred to by the celestial directions NW, SW, SE and NE. Figure 3 illustrates an example of a decomposed geometry and the corresponding tree data structure.

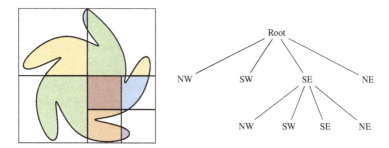

Fig. 3 Cell decomposition and tree data structure

The uniform structure ensures a fast search for neighbouring quadtree cells, i.e., such cells that share an edge. It is obvious that each interface of the resulting multi-patch lies on such an edge. Furthermore, the concept is easily generalizable to higher dimensions. For example, an octree stores spatial data and is characterized by each inner node having eight children. A far more detailed description of quadtrees can be found in [6, Chap. 14].

In the upcoming sections we will introduce the crucial steps for decomposing a domain by a recursive quadtree algorithm.

3.1 Star-Shapedness

Let the domain Ω be defined by a boundary B-spline-parametrization γ as in (3), based on a knot vector Ψ, a polygonal degree p and the (weighted) control polygon $\mathscr{P}$. These three elements are stored in arrays in case the domain has multiple boundaries, where for each index it is specified whether the entries belong to a hole (void) or an outer boundary. This can either be done explicitly by a boolean flag or implicitly by conventions asking the user for example to parametrize outer boundaries counter-clockwise and holes clockwise, or simply restricting the input to one outer boundary such that connected components are handled separately.

The concept of visibility is elementary for the theory of star domains. Therefore, consider two points $x, y \in \Omega$. Now y is said to be *visible* from x within Ω if the direct line segment $\overline{xy}$ lies completely in Ω. In this case x is said to *see* (or *cover*) y. The set of all visible points from x in Ω is denoted by $\mathrm{vis}_\Omega(x)$. The set of all points x_0 overseeing the whole domain, i.e. with $\mathrm{vis}_\Omega(x_0) = \Omega$, is called kernel and denoted as $\ker \Omega$. Elements from the kernel are referred to as *star-centers*. A domain is called *star-shaped* (or *star-convex* or simply a *star domain*) if it possesses a non-empty kernel.

Apparently, visibility is based on straight lines L and their property of entering and leaving the domain. Therefore, we distinguish between *regular* intersections (denoted by $\partial \Omega \cap L$) and *proper* intersections (denoted by $\Omega \hat{\cap} L$) and define

$$\Omega \hat{\cap} L := \{ y \in \partial \Omega \cap L \mid \forall \epsilon > 0 \, \exists x \in L \setminus \Omega : \; \|y - x\| < \epsilon \}.$$

Hence, proper intersections are exactly the points where L enters and leaves Ω. Figure 4 displays an example setting with proper intersection points y_i and in contrast x which is only a regular intersection point.

Lemma 1 *Let Ω be a bounded and closed domain and $x_0 \in \Omega$. Then x_0 is a star-center of Ω if and only if each line through x_0 has exactly two proper intersections with Ω.*

Fig. 4 Proper intersections y_1, y_2, y_3 of line and domain and regular intersection point x

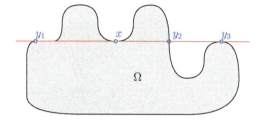

Proof "$\Rightarrow$" Let $x_0 \in \ker \Omega \subseteq \Omega$. The boundary $\partial \Omega$ is trivially contained in the visible space of x_0. Furthermore, let L be a straight line going through x_0. Since $x_0 \in \Omega$ and Ω is bounded, there must be at least two proper intersection points of L and Ω.

Assume that there were at least three proper intersection points y_1, y_2, y_3, labelled such that y_2 lies on the line segment $\overline{y_1 y_3}$. Now, the goal is to force y_2 to be equal to one of the other intersection points. Since x_0 covers the boundary and hence all three intersection points, both line segments $\overline{x_0 y_1} \subset L$ and $\overline{x_0 y_3} \subset L$ are contained in Ω. This directly yields that the segment $\overline{y_1 y_3} \subset L$ lies completely in Ω. Thus, $y_2 \in \{y_1, y_3\}$.

"$\Leftarrow$" Let $x_0 \in \Omega$ such that each line through x_0 has exactly two proper intersections with Ω. Choose $z \in \Omega \setminus \{x_0\}$ arbitrarily and let L_z be the unique line through x_0 and z. By assumption it holds that

$$x_0, z \in \Omega \cap L_z = \overline{y_1 y_2}$$

with $\{y_1, y_2\} = L_z \hat{\cap} \Omega$. Therefore, the line segment $\overline{x_0 z}$ lies completely in Ω. By arbitrary choice of z, x_0 is a star-center. $\square$

Note that the kernel of a polygon is a polygon again and given as the intersection of halfspaces induced by the polygon edges as depicted in Fig. 5a. However, it is quite hard to compute the kernel of a continuous domain. The following theorem describes the connection of control polygon and boundary curve with respect to star-shapedness. This will be exploited by the recursion criterion.

Theorem 1 *Star-centers of the (unweighted) control polygon $\mathscr{P}$ are inherited by the domain. More precisely it holds*

$$\ker \mathscr{P} \cap \Omega \subseteq \ker \Omega. \qquad (9)$$

Proof Without loss of generality assume there is $x_0 \in \ker \mathscr{P} \cap \Omega \neq \emptyset$. Then by the characteristics for kernels (Lemma 1) each line through x_0 properly intersects $\partial \mathscr{P}$ exactly twice. By the variation diminishing property, the image of the B-spline curve $\partial \Omega$ is intersected at most twice, as well. Since $x_0 \in \Omega$ there are at least two intersection points of L and $\partial \Omega$. All together each line through x_0 properly intersects the boundary exactly twice and $x_0 \in \ker \Omega$ again by Lemma 1. $\square$

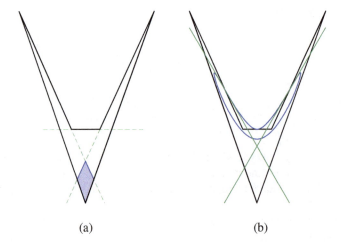

Fig. 5 V-shaped example domain. (**a**) Kernel computation for polygons. (**b**) No star-shaped domain

Figure 5a illustrates the process of kernel computation. The kernel of the V-shaped domain is depicted as the blue area. Note that a star-shaped control polygon does not guarantee star-shapedness of the corresponding B-spline domain. In Fig. 5b a cubic uniform rational B-spline curve based on the V-shaped polygon is painted in blue. Two tangents at the steepest positions show that the domain bounded by this curve is not a star domain.

3.2 Recursion Scheme

Since the quadtree creation is based on recursive refinement, there is need for a termination criterion that notices when a square region is star-shaped. For this purpose first search for the presence of holes, as these prevent star-convexity. Afterwards compute the kernel of the control polygon.

In case the control polygon is star-shaped, the intersection of the kernel of the polygon and the actual domain is constructed. For efficiency, it suffices to check if any edge of the kernel intersects with the boundary of the domain and if one of the vertices lies inside. If an element of $\ker \mathscr{P} \cap \Omega$ is found, the algorithm can propose this as a star-center and return. Otherwise, further refinement is initiated.

3.3 Intersection of Rays and Splines

The core mechanic behind continuous quadtree refinement of B-spline curves is intersecting such with rays—especially with horizontal or vertical lines. For this we interpret a line G as a one-dimensional hyperplane given by its normal vector g_n and a translation g_o. Then the task of intersecting the line with a B-spline curve γ is equivalent to finding $\eta \in [0, 1]$ such that the non-linear equation

$$\langle \gamma(\eta) - g_o, g_n \rangle = 0 \qquad (10)$$

holds. A solution can be computed by numerical root finding techniques as, e.g., Newton's method. Based on the local modification scheme of B-spline curves, intersecting the ray with the control polygon in advance yields a good initial guess for iterative solvers.

3.4 Quadtree-Based Refinement Algorithm

Altogether these pieces form the final refinement algorithm: Given a series of boundary curves, start by checking for star-convexity as described in Sect. 3.2. Each boundary curve describing a star-shaped patch is added to the tree data structure at the current node. All other boundaries are kept for recursion, which is applied by computing the centroid of the quadtree cell and intersecting the curves with the horizontal and vertical lines through the centroid.

The resulting curve segments are concatenated with line segments following the interfaces (Fig. 6). This yields a list of closed boundary curves for each quadrant again. Then, the quadtree algorithm is called recursively on each quadrant yielding the child nodes for the currently handled node.

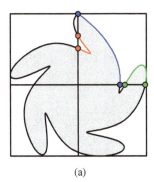
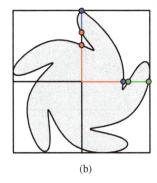

(a)   (b)

Fig. 6 Illustration of the merging step in north-eastern quadrant. (**a**) Pairing of intersection points. (**b**) Pairing of curve segments

Note that it is meaningful not to merge the interfaces with the spline-curves yet, but to keep them as line segments instead. This not only increases performance of the intersection process in coming recursion steps, but also marks them as an interface.

3.5 Setup of Interfaces for Numerical Treatment

The computation of suitable interfaces is crucial for the application of a numerical method on the resulting multi-patch geometry. The first step is to set up conformal interface parametrizations such that condition (8b) can be ensured easily by the numerical solver. Hence, for each boundary curve γ resulting from the decomposition find all the curves (partly) sharing an interface with γ. By construction those curves lie in the neighbouring quadtree cells which can be found very quickly.

Let $\Omega^{(k)}$, $\Omega^{(l)}$ be patches resulting from the quadtree refinement (partly) sharing an edge $\overline{z_1 z_2}$. If both are located at the same depth within the quadtree, they completely share the edge which can be parametrized trivially as p-th degree B-spline curves with

$$\Psi_{edge} = \{\underbrace{0,\ldots,0}_{p+1 \text{ times}}, \underbrace{1,\ldots,1}_{p+1 \text{ times}}\}, \qquad \mathscr{P}_{edge} = \{\frac{r}{p}z_2 + \frac{p-r}{p}z_1 | r = 0,\ldots,p\}$$

where p is the maximum degree of the boundary curves of $\Omega^{(k)}$ and $\Omega^{(l)}$. The setting is depicted in Fig. 7. Note that the order of control points may be reversed for $\Omega^{(l)}$. With this setup we can merge the spline curve of the edge with the associated boundary parts.

In case $\Omega^{(l)}$ belongs to a descendant of the cell $\mathscr{B}$ neighbouring $\Omega^{(k)}$, basically the same procedure can be applied—but only for the shared part of the interface. However, the remaining sections are handled regarding other descendants of $\mathscr{B}$.

In order to describe the interface $\Gamma^{(k,l)}$ the indices of matching control points from $\mathscr{P}_{edge}$ within the total boundary curves of $\Omega^{(k)}$ and $\Omega^{(l)}$ are stored. Additionally, for each patch all indices of control points belonging to an interface are stored in a list, where the points belonging to both an interface and the boundary $\partial \Omega^{(k)}$ are marked individually.

Fig. 7 Example of the interface parametrization

3.6 Benefits, Termination and Difficulties

Our quadtree approach comes up with several very nice properties. As already described at the start of this section, we end up with a tree data structure which implicitly contains the interface description. The further effort to set up the multipatch description is very low. Moreover, the regular and rectangular refinement might pose advantages for some problems.

The recursion scheme derived in Sect. 3.2 provides robustness with respect to star-shapedness, as the algorithm only terminates when the currently regarded subdomain is star-shaped. However, the scheme might refine although the domain is already star-shaped, as the criterion is only sufficient, but not necessary.

Termination of the algorithm is given by the variation diminishing property of B-splines. They follow their control polygon smoothly and knot insertion will make the polygon adapt to the curve. Therefore, the variation within the boundary is predefined at the start of the algorithm and we terminate as soon as the diameter of the present quadtree cell falls below $\alpha \Delta \mathscr{P}$ with

$$\Delta \mathscr{P} := \min_{P_i \neq P_j} \| P_i - P_j \|$$

and $\alpha \in \mathbb{R}$ is a constant independent of the B-spline curve.

A disadvantage of the quadtree is that its strategy does not adapt to the domain. We are always splitting along horizontal and vertical lines through the center points of the cells. Figure 8 displays an example domain for which the quadtree needs unnecessarily many refinement steps. Thus, the result is composed of a huge amount of patches with $\mathscr{C}^0$ interfaces whereas only three such borders are necessary. This can be avoided by mimicking the pivoting of kD-trees (cf. [6, Chap. 5]). These possess analogous capabilities of data storage as the quadtree, however, the kD-tree-refinement is based on event points computed either beforehand or at runtime.

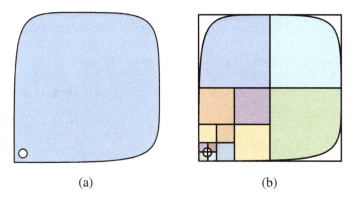

(a)   (b)

Fig. 8 Degenerated example for the quadtree algorithm. (**a**) Domain. (**b**) Quadtree decomposition

When equipped with a suiting event searching strategy the kD-tree approach will certainly come up with a decomposition consisting of fewer patches than the quadtree algorithm. However, we then loose the possibility of fast interface setup due to complicated neighbour search.

The biggest drawback of the quadtree algorithm is the intersection and splitting process in higher dimensions. For example, in 3D the analytical solution to (10) is a NURBS curve itself. Furthermore, the boundary surface must be split along a curve which in most cases is not an isoparametric curve. Hence, the need for such a splitting process arises.

4 Art Gallery Decomposition

The second approach for refinement into star-shaped pieces is based on the local modification scheme of B-splines, after which the curve follows the control polygon smoothly. Hence, the polygon is broken down first and the decomposition is transferred to the domain afterwards.

The task of dividing a polygon into star-convex pieces is formally known as "Art Gallery Problem" and well-understood [21]. For illustration purposes we apply the very simple method of Fisk, equipped with some graph-theoretical background in the covering step. The notation is adopted from [17], and the concepts linear duality and complementary slackness stem from [23, Chap. 7].

4.1 Fisk's Algorithm

The main idea of Fisk's method is to triangulate the underlying polygon and to merge suitable adjacent triangles together afterwards. The single steps of the algorithm are illustrated in Fig. 9. The result of the triangulation is a decomposition of the polygon into pieces consisting of three vertices and edges, respectively. The construct can be interpreted as an undirected graph with vertices and edges being defined analogously. It is obvious that each vertex can oversee all incident triangles.

The main task when merging the triangles is to find a set of vertices such that all triangles—i.e. cliques of size three—are covered. Therefore, we call this problem minimum three-cliques cover (M3CC). The corresponding integer program for the binary variable s_v that marks the vertices reads

$$\text{minimize} \sum_{v \in V} s_v$$

$$\text{s.t.} \sum_{v \in \Delta} s_v \geq 1 \quad \forall \Delta \in T_3,$$

$$s_v \in \{0, 1\} \quad \forall v \in V,$$

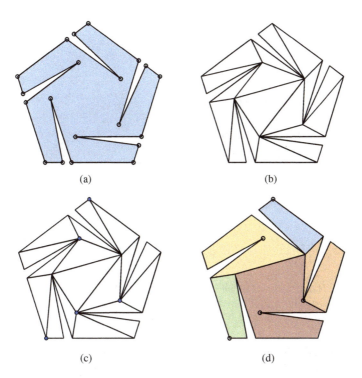

Fig. 9 Fisk's method applied on a rotor-shaped polygon. (**a**) Rotor polygon. (**b**) Triangulation. (**c**) Triangle cover. (**d**) Decomposed polygon

where T_3 is the set of triangles and V is the set of vertices within the triangulation. In the original form Fisk approximates M3CC by a vertex colouring. However, this fails if the polygon contains a hole or if we are working in higher dimensions. Thus, we employ general solution methods to this problem.

Though several approaches as branch and bound or cutting planes [23, Chaps. 23–24] have been developed, solving general integer programs optimally is *NP*-complete. However, simulations on test domains do not show exponential scaling of the runtime with respect to the number of control points indicating some beneficial structure of our specific problem. Nevertheless, discrete optimisation has some standard heuristics at hand as for example Greedy-type algorithms which run in polynomial time and approximate the optimal solution very well. The interested reader is referred to [19, Chap. 12] for a detailed overview of approximative approaches for integer problems. After having solved the M3CC, vertices are taken iteratively from the solution cover and all unhandled adjacent triangles are merged together, which yields a decomposition of the polygon.

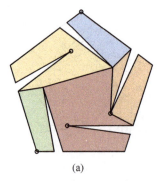 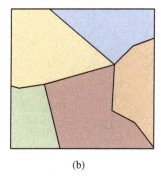

Fig. 10 Extension of tessellation to bounding box. (**a**) Polygon tesselation. (**b**) Voronoi diagram

4.2 Voronoi Diagrams for Exterior Visibility

Now that we have a decomposition of the control polygon, we want to transfer this refinement to the actual domain. Our approach to this problem is motivated by Voronoi diagrams (cf. [6, Chap. 7]) as a tessellation of the axis aligned bounding box (Fig. 10).

Algorithmically, the main task of the algorithm is to compute the multipatch-interfaces based on the interfaces of the polygon decomposition. The final set of interfaces consists of three different types of interfaces:

- interfaces of the polygon decomposition fully contained by Ω
- segments of the interfaces of the decomposition of the polygon intersecting $\partial \Omega$
- additional interfaces for exterior visibility where patches meet

The first two classes arise directly from the polygon tessellation and can be distinguished by intersecting interfaces with the boundary curve as described in Sect. 3.3 and asserting that the intersection point lies on the interface segment. The setup of the third class is more involved, though.

First, all control points belonging to more than one patch and lying inside of Ω must be found. The first criterion is met exactly by the end points of interfaces. Therefore, the list of those vertices can be created as a by-product of handling the first two interface classes. According to the Voronoi diagram, a new interface arises starting at those points, moving in the direction of the bisector of the polygon edges and ending where it intersects the boundary curve γ.

As there might arise cases where this strategy induces non-convex vertices—i.e. vertices with greater inner than outer angle between incident edges—in the concerned patches, we consider the incident interfaces first. In case there are such on both sides of the bisector of the edge, any choice will yield convex vertices for the adjacent patches. Otherwise we check whether an interface can be elongated such that it intersects the boundary curve before meeting with the control polygon again. If this is not possible, we choose the bisector nevertheless.

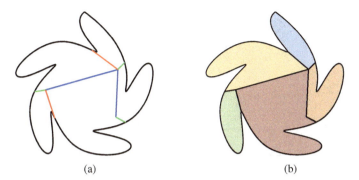

Fig. 11 Final decomposition via art gallery method. (**a**) Interface graph diagram. (**b**) Domain tessellation

The three types of interfaces equipped with their end points form a (not necessarily connected) undirected graph. We add edges between those points lying on and directly connected via the boundary curve and mark them as such. The resulting "interface graph diagram" directly yields the domain decomposition as it is depicted in Fig. 11. The three interface classes are coloured in blue, green and red, respectively.

All the interface parametrization, the splitting and merging of boundary parts and data stored for the interfaces is analogous to the quadtree decomposition from Sect. 3.5. However, this time each edge describing an interface is completely shared by two patches.

4.3 Robustness and Higher Dimensions

In contrast to the quadtree algorithm our art gallery approach tries to minimize the number of patches adaptively with respect to the control polygon. Yet again, knot insertion before applying the decomposition algorithm does not change the geometry at all, but will impact the refinement and the number of patches.

Moreover, the art gallery approach has several problems with star-convex and "nearly" star-convex domains. Considering the rotor example domain, our plain standard art gallery approach with Voronoi extension yields two patches which are not star-shaped (coloured in yellow and red). This can be easily remedied by using polygon edges instead of bisectors for the third class of interfaces (Fig. 12). However, an automatic recognition scheme for these cases at runtime is still missing. Another example in which our approach fails is the V-shaped domain from Fig. 5. Here, the control polygon is already star-shaped and thus does not require a decomposition by the art gallery method, but the domain itself does not need to be star-shaped. Furthermore, Fig. 13a shows the epitome of star domains. Still the art gallery approach decomposes it into smaller blocks.

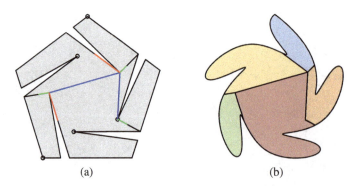

Fig. 12 Art gallery decomposition of rotor domain. (**a**) Adapted interface choice. (**b**) Adapted decomposition

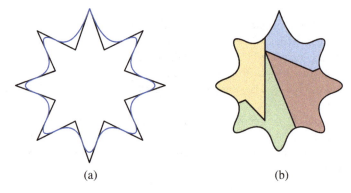

Fig. 13 Failure of the art gallery approach on a star domain. The decomposition process starts despite there is no need for it. (**a**) Star polygon and domain boundary. (**b**) Art gallery decomposition

Whereas the splitting process in higher dimensions is analogously difficult as in the quadtree case, the rest of the algorithm is perfectly suitable for higher dimensions. In 3D for example, a tetrahedralisation replaces the triangulation. The cover problem generalizes trivially to higher dimensions as MdCC with d being the dimension plus one. Another benefit is that the elements of the approach—namely triangulation/tetrahedralization, integer programming and Voronoi diagrams—are already well-studied and there are very fast solution algorithms. Moreover, the ingredients, as for example the list of triangles or the first two interface classes in the transfer step, are computed as by-product.

5 Numerical Simulations

In this section we apply the decomposition algorithms from Sects. 3 and 4 to several benchmark domains. First we decompose the rotor which we already used for illustration purposes and the Yeti footprint from [14] in order to solve the Poisson equation (5). Afterwards we illustrate the convergence of our numerical method on an annulus-shaped domain. The patches resulting from refinement, originally given in boundary description, are then parametrized as described in Sect. 2.2 by means of the scaled-boundary technique. For the computation of star-centers we approximate the boundary curve by a polygonal chain and choose the centroid of its kernel.

All methods have been implemented in Matlab. We solve the partial differential equation by an extension of the ISOGAT package [26] which handles multipatch geometries with matching interfaces $\Gamma^{(k,l)}$. The continuity across interfaces is enforced by means of corresponding constraints that are used to eliminate the superfluous control points or degrees of freedom, respectively, along the interface. The same elimination process is applied to the multiple control point in the scaling center of each block, see [1] for the details. In the same reference [1], it is also shown that the combination of a scaled boundary parametrization and a standard Galerkin-bases IGA code, as we apply it here, is computationally equivalent to the SB-IGA approach of [15] for linear elliptic problems.

5.1 The Rotor with Five Wings

First we consider the rotor geometry with five wings (Fig. 14). On this domain we solve the Poisson equation (5) with constant right hand side $f = 1$ and zero Dirichlet boundary conditions.

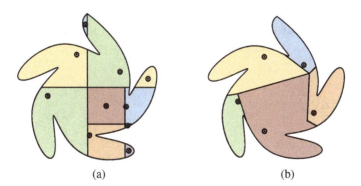

Fig. 14 Decompositions of the rotor with marked star-centers [number of patches]. (**a**) Quadtree decomposition [10]. (**b**) Art gallery decomposition [5]

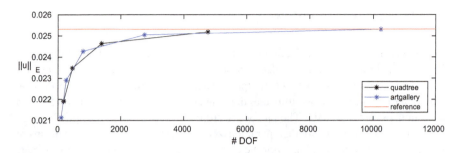

Fig. 15 Convergence of the solution of the rotor geometry in energy norm

(a)   (b)

Fig. 16 Simulation results for the benchmark geometries. (**a**) Solution on rotor. (**b**) Solution on yeti footprint

Figure 15 illustrates the convergence in energy norm with respect to h-refinement. The black and blue lines stand for the quadtree and art gallery approach, respectively. The red line indicates the reference value stemming from the standard Galerkin approach with cubic B-splines in radial direction after three h-refinements. As bottom line, we observe that both decomposition approaches yield comparable results. The three-dimensional surface plot Fig. 16a displays the numerical solution based on the art gallery decomposition and the scaled boundary approach from Sect. 2.3 after one h-refinement, with quadratic B-splines for each patch.

5.2 The Yeti Footprint

The geometrically more interesting example is the Yeti footprint [14]. Figure 17 displays the decompositions that our methods have constructed for this geometry with four holes and multiple boundaries. Again we solve (5) with constant right hand side $f = 1$ and zero Dirichlet boundary conditions and consider the convergence in energy norm.

The surface plot Fig. 16b displays the numerical solution based on the quadtree decomposition and the scaled boundary approach from Sect. 2.3 after one h-

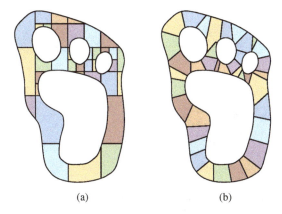

Fig. 17 Decompositions of the Yeti footprint [number of patches]. (**a**) Quadtree decomposition [39]. (**b**) Art gallery decomposition [38]

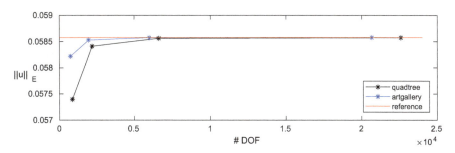

Fig. 18 Convergence of Yeti's solution in energy norm

refinement. Figure 18 illustrates the convergence in energy norm. Note that for this example, the decomposition by the art gallery approach yields faster convergence. This can be explained by several tiny patches that are generated in the quadtree algorithm and that lead to a non-balanced multipatch structure.

5.3 Convergence on an Annulus

In order to better assess the convergence behavior of the resulting discretizations, we add another rather simple example. Let the domain Ω be given by two circular boundaries with radius 1 and 2, i.e.

$$\Omega = \{(x, y) \in \mathbb{R}^2 | 1 \leq x^2 + y^2 \leq 4\},$$

where the two boundaries are given by quadratic NURBS curves starting and ending at the positive part of the x-axis.

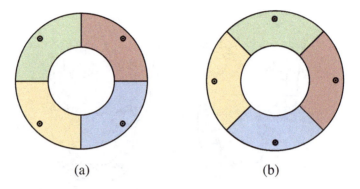

Fig. 19 Decompositions of the annulus domain with marked star-centers [number of patches]. (**a**) Quadtree decomposition [4]. (**b**) Art gallery decomposition [4]

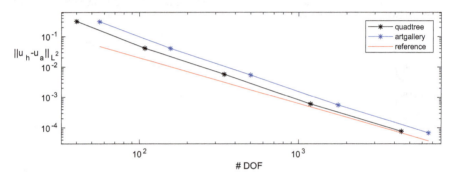

Fig. 20 Convergence of the numerical solution in L^2-norm

The results of our decomposition methods are displayed in Fig. 19. Note that both approaches yield basically the same decomposition of the real domain with differences only in parametrization.

The benchmark problem is the Poisson equation with zero Dirichlet boundary conditions and right-hand side

$$f = 4\beta^2 R \sin(\beta(R-1)) - 4\beta \cos(\beta(R-1)),$$

where $R = x^2 + y^2$ and $\beta = \frac{\pi}{3}$ for enhanced readability. It is easily verified that the analytic solution is a radial symmetric sine wave given by

$$u_a(x,y) = \sin\left(\frac{\pi}{3}(x^2 + y^2 - 1)\right).$$

Figure 20 displays the convergence of our approaches compared to a reference line representing $\mathcal{O}(\#\text{DOF}^{-\frac{3}{2}}) = \mathcal{O}(h^3)$. Since quadratic B-splines are used as basic discretization for each patch, the convergence behavior of the multipatch simulation is as expected, with the quadtree approach performing somewhat better. Note that the art gallery decomposition introduces new control points around the interfaces by knot insertion, whereas the quadtree algorithm draws interfaces exactly at the images of knots with multiplicity equal to 2, which results in less degrees of freedom and thus is a possible explanation for the slightly better performance.

6 Conclusions

We have discussed the decomposition of a planar domain into star-shaped blocks, motivated by the idea of solving PDEs by means of the scaled boundary isogeometric analysis. Both approaches considered, the quadtree decomposition and the art gallery decomposition, turned out to be suitable for this purpose, with each having its merits and its drawbacks.

The quadtree algorithm is fast and provides a tree data structure which implicitly contains the interface description, which means that the further effort to set up the multipatch description is very low. Moreover, the recursion scheme derived in Sect. 3.2 provides robustness with respect to star-shapedness, but the scheme might refine although the domain is already star-shaped. The biggest drawback of the quadtree algorithm is the intersection and splitting process if one tries to tackle 3D geometries.

In contrast to the quadtree algorithm the art gallery approach tries to minimize the number of patches. The elements of the approach—namely triangulation, integer programming and Voronoi diagram—are already well-studied and there exist very fast algorithms for these steps. Nevertheless, if we want to extend this approach to 3D, the splitting process is analogously difficult as in the quadtree case.

Computationally, the performance depends strongly on the chosen example geometry, and we cannot give a general recommendation which approach should be preferred.

Acknowledgments We are grateful to Sven Klinkel for fruitful discussions on the various possibilities to decompose a not star-shaped domain in order to apply SB-IGA. Moreover, we acknowledge the support by the European Union under grant no. 678727 (Project MOTOR) and by the German Research Council (DFG) under grant no. SI 756/5-1 (Project YASON).

References

1. C. Arioli, A. Shamanskiy, S. Klinkel, and B. Simeon. Scaled boundary parametrizations in isogeometric analysis. *Comput. Meth. Appl. Mech. Engrg*, 349:576–594, 2019.
2. M. Chasapi and S. Klinkel. A scaled boundary isogeometric formulation for the elasto-plastic analysis of solids in boundary representation. *Comput. Meth. Appl. Mech. Engrg.*, 333:475–496, 2018.

3. L. Chen, W. Dornisch, and S. Klinkel. Hybrid collocation-Galerkin approach for the analysis of surface represented 3D-solids employing SB-FEM. *Comput. Meth. Appl. Mech. Engrg.*, 295:268–289, 2015.
4. L. Chen, B. Simeon, and S. Klinkel. A NURBS based Galerkin approach for the analysis of solids in boundary representation. *Comput. Meth. Appl. Mech. Engrg.*, 305:777–805, 2016.
5. J. A. Cottrell, T. JR Hughes, and Y. Bazilevs. *Isogeometric analysis: toward integration of CAD and FEA*. John Wiley & Sons, 2009.
6. M. de Berg, O. Cheong, M. van Kreveld, and M.H. Overmars. *Computational Geometry: Algorithms and Applications*. Springer-Verlag TELOS, Santa Clara, CA, USA, 3rd edition, 2008.
7. C. de Boor. *A Practical Guide to Splines*. Springer, 1978.
8. H. Gravenkamp, A. Saputra, C. Song, and C. Birk. Efficient wave propagation simulation on quadtree meshes using SBFEM with reduced modal basis. *International Journal for Numerical Methods in Engineering*, 110:1119–1141, 06 2017.
9. J. Gravesen, A. Evgrafov, D.-M. Nguyen, and P. Nørtoft. Planar parametrization in isogeometric analysis. In M. Floater, T. Lyche, M.-L. Mazure, K. Mørken, and L. L. Schumaker, editors, *Mathematical Methods for Curves and Surfaces: 8th International Conference, MMCS 2012, Oslo, Norway, June 28 – July 3, 2012, Revised Selected Papers*, pages 189–212. Springer Berlin Heidelberg, 2014.
10. C. Hofer and U. Langer. Dual-primal isogeometric tearing and interconnecting solvers for multipatch dG-IgA equations. *Comput. Meth. Appl. Mech. Engrg*, 316:2–21, 2017. Special Issue on Isogeometric Analysis: Progress and Challenges.
11. T. J. R. Hughes, J. A. Cottrell, and Y. Bazilevs. Isogeometric analysis: CAD, finite elements, NURBS, exact geometry and mesh refinement. *Comput. Meth. Appl. Mech. Engrg.*, 194:4135–4195, 2005.
12. S. A. Joshi, Y. Rao, B. R. Sundar, and R. Muthuganapathy. On the visibility locations for continuous curves. *Computers & Graphics*, 66:34–44, 2017.
13. M. Kapl, G. Sangalli, and T. Takacs. Construction of Analysis-suitable G1 planar Multi-patch Parameterizations. *Computer-Aided Design*, 97:41–55, 2018.
14. S. Kleiss, C. Pechstein, B. Jüttler, and S. Tomar. Ieti – isogeometric tearing and interconnecting. *Comput. Meth. Appl. Mech. Engrg*, 247–248:201–215, 2012.
15. S. Klinkel, L. Chen, and W. Dornisch. A NURBS based hybrid collocation-Galerkin method for the analysis of boundary represented solids. *Comput. Meth. Appl. Mech. Engrg.*, 284:689–711, 2015.
16. S. Klinkel and R. Reichel. A finite element formulation in boundary representation for the analysis of nonlinear problems in solid mechanics. *Comput. Meth. Appl. Mech. Engrg.*, 347:295–315, 2019.
17. S. O. Krumke and H. Noltemeier. *Graphentheoretische Konzepte und Algorithmen*. Leitfäden der Informatik. Vieweg+Teubner Verlag, 2012.
18. U. Langer, A. Mantzaflaris, S. E Moore, and I. Toulopoulos. Multipatch discontinuous galerkin isogeometric analysis. In *Isogeometric Analysis and Applications 2014*, pages 1–32. Springer, 2015.
19. K. Mehlhorn and P. Sanders. *Algorithms and Data Structures: The Basic Toolbox*. Springer, 2008.
20. S. Natarajan, J. C. Wang, C. Song, and C. Birk. Isogeometric analysis enhanced by the scaled boundary finite element method. *Comput. Meth. Appl. Mech. Engrg.*, 283:733–762, 2015.
21. J. O'Rourke. *Art Gallery Theorems and Algorithms*. Oxford University Press, Inc., New York, NY, USA, 1987.
22. L. Piegl and W. Tiller. *The NURBS book*. Monographs in visual communications. Springer, 1997.
23. A. Schrijver. *Theory of Linear and Integer Programming*. John Wiley & Sons, Inc., New York, NY, USA, 1986.

24. C. Song and J. P. Wolf. The scaled boundary finite-element method-alias consistent infinitesimal finite-element cell method-for elastodynamics. *Comput. Methods Appl. Mech. Engrg.*, 147:329–355, 1997.
25. C. Song and J. P. Wolf. The scaled boundary finite-element method–a primer: solution procedures. *Comput. Struct.*, 78(1):211–225, 2000.
26. A.-V. Vuong, Ch. Heinrich, and B. Simeon. Isogat: A 2d tutorial matlab code for isogeometric analysis. *Computer Aided Geometric Design*, 27(8):644–655, 2010. Advances in Applied Geometry.
27. G. Xu, B. Mourrain, R. Duvigneau, and A. Galligo. Parametrization of computational domain in isogeometric analysis: methods and comparison. *Comput. Meth. Appl. Mech. Engrg.*, 200(23–24):2021–2031, 2011.

Approximation Power of C^1-Smooth Isogeometric Splines on Volumetric Two-Patch Domains

Katharina Birner, Bert Jüttler, and Angelos Mantzaflaris

Abstract Bases and dimensions of trivariate spline functions possessing first order geometric continuity on two-patch domains were studied in Birner et al. (Graph Mod 99:46–56, 2018). It was shown that the properties of the spline space depend strongly on the type of the gluing data that is used to specify the relation between the partial derivatives along the interface between the patches. Locally supported bases were shown to exist for trilinear geometric gluing data (that corresponds to piecewise trilinear domain parameterizations) and sufficiently high degree. The present paper is devoted to the approximation properties of these spline functions.

We recall the construction of the basis functions and show how to compute them efficiently. In contrast to the results in Birner et al. (Graph Mod 99:46–56, 2018), which relied on exact arithmetic operations in the field of rational numbers, we evaluate the coefficients by computations with standard floating point numbers. We then perform numerical experiments with L^2-projection in order to explore the approximation power of the resulting spline functions. Despite the existence of locally supported bases, we observe a reduction of the approximation order for low degrees, and we provide a theoretical explanation for this locking.

1 Introduction

The framework of Isogeometric Analysis [7] facilitates numerical simulation with high-order partial differential equations, since it supports C^r-smooth discretizations with $r > 0$. For $r = 1$, these are especially useful when considering PDEs of order four, such as the Cahn-Hilliard equation [8], shells [2], and the biharmonic equation [1].

K. Birner · B. Jüttler (✉)
Institute of Applied Geometry, Johannes Kepler University, Linz, Austria
e-mail: katharina.birner@jku.at; bert.juettler@jku.at

A. Mantzaflaris
Inria Sophia Antipolis – Méditerranée, Université Côte d'Azur, Nice, France

While the construction of C^1-smooth spline functions on single patches is straightforward, the extension to multi-patch domains, which are needed to describe more complex domains, requires the notion of geometric continuity [16]. More precisely, C^r-smoothness ($r \geq 0$) of an isogeometric spline function is implied by G^r-smoothness (geometric continuity of order r) of the associated graph surface [9]. This result has widely been used to construct C^1-smooth spline spaces on planar domains.

Smooth approximations over unstructured quadrilateral meshes were considered in [3]. The construction of geometrically continuous splines on arbitrary topologies was studied in [14]. Bases and dimensions of the space of C^1-smooth isogeometric functions for bilinearly parameterized domains were explored in [10]. Some of these results have been extended to C^2-smooth splines [12].

The numerical examples presented in these publications indicate optimal approximation power for combinations of sufficiently high degrees with certain classes of gluing data. In particular, the generalization of parameterizations with bilinear gluing data to the more general class of analysis-suitable parameterizations, which appears to preserve the optimal approximation properties, was presented in [6, 11].

The extension to trivariate domains was studied in [4, 5, 15]. The domains considered in [15] are obtained via sweeping, which restricts the available topologies. In [4], we studied the space of globally C^1-smooth splines on a two-patch domain, which is topologically equivalent to two cuboids. We considered different types of gluing data and presented the associated dimension formulas. Moreover, we showed how to construct a basis and identified those types of gluing that yield locally supported basis functions indicating good approximation properties.

The space of C^1-smooth isogeometric functions for trilinearly parameterized two-patch domains was further studied in [5]. Explicit representations of the locally supported basis functions were presented and the numerically obtained dimension formula from [4] was theoretically verified.

The present paper extends these existing results. In Sect. 2, we recall the notion of the glued spline space $\mathfrak{G}_D$, which characterizes the space of C^1-smooth isogeometric functions $\mathcal{V}_F$. Based on these preparations, Sect. 3 considers the coefficient patterns of trilinear geometric gluing data, which was found to be promising for good approximation power in [4], in further detail. These patterns allow us to efficiently compute a basis of the space $\mathcal{V}_F$ for this type of gluing data. Finally, in Sect. 4 we numerically analyze the approximation power of the basis via L^2-fitting. We conclude the paper in Sect. 5.

2 Preliminaries

We consider two subdomains $\Omega^{(1)}$ and $\Omega^{(2)}$, both topologically equivalent to a cube, which form the two-patch geometry $\Omega = \Omega^{(1)} \cup \Omega^{(2)}$. Let $\mathcal{S}_k^p$ denote the space of spline functions on $[0, 1]$ of degree p with k uniformly distributed inner knots of

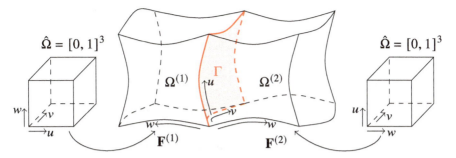

Fig. 1 Parameterizations of two volumetric subdomains $\Omega^{(1)}$ and $\Omega^{(2)}$ joined at the interface Γ

multiplicity $p - 1$. We use it to define the tensor-product space

$$\mathscr{P} = \mathscr{S}_k^p \otimes \mathscr{S}_k^p \otimes \mathscr{S}_k^p,$$

which consists of C^1-smooth trivariate spline functions.

The two subdomains are described by parametric representations $\boldsymbol{F}^{(i)} : \hat{\Omega} = [0, 1]^3 \to \Omega^{(i)}$, $i = 1, 2$ with coordinate functions from $\mathscr{P}$, see Fig. 1. These define the two-patch geometry mapping

$$\boldsymbol{F} = (\boldsymbol{F}^{(1)}, \boldsymbol{F}^{(2)}) \in \mathscr{P}^3 \times \mathscr{P}^3.$$

We assume that the two patches meet with C^0-smoothness

$$\boldsymbol{F}^{(1)}(u, v, 0) = \boldsymbol{F}^{(2)}(u, v, 0), \quad u, v \in [0, 1]$$

at the common interface $\Gamma = [0, 1]^2 \times \{0\}$. In this paper, we explore the space

$$\mathscr{V}_F = [(\mathscr{P} \times \mathscr{P}) \circ \boldsymbol{F}^{-1}] \cap C^1(\Omega^{(1)} \cup \Omega^{(2)})$$

of C^1-smooth isogeometric functions on Ω.

The elements of the pairs $f = (f^{(1)}, f^{(2)}) \in \mathscr{P} \times \mathscr{P}$ possess representations

$$f^{(i)} = \sum_{k=0}^n b_k^{(i)} N_k(u, v, w), \quad i = 1, 2,$$

with real coefficients $b_k^{(i)}$, where the symbols N_k denote the tensor-product B-splines that span the space $\mathscr{P}$. The glued spline space $\mathfrak{G}_D$ is a subspace of $\mathscr{P} \times \mathscr{P}$, which was introduced in [4]. For given gluing data $D = (\beta, \gamma, \alpha^{(1)}, \alpha^{(2)})$, which is a quadruple of four bivariate polynomials, it consists of all functions whose coefficients satisfy

$$
\begin{aligned}
0 = &\sum_{k=0}^{n} b_k^{(1)} N_k(u,v,0) - b_k^{(2)} N_k(u,v,0) \quad \text{and} \\
0 = &\sum_{k=0}^{n} b_k^{(1)} \Big(\beta(u,v)\,(\partial_u N_k)(u,v,0) - \gamma(u,v)\,(\partial_v N_k)(u,v,0) \\
& + \alpha^{(1)}(u,v)\,(\partial_w N_k)(u,v,0) \Big) - b_k^{(2)} \alpha^{(2)}(u,v)\,(\partial_w N_k)(u,v,0).
\end{aligned}
\tag{1}
$$

Bases and dimensions of the glued spline space $\mathfrak{G}_D$ for different types of gluing data were studied in [4]. Furthermore it was observed that

$$\mathscr{V}_F = \mathfrak{G}_D \circ \mathbf{F}^{-1},$$

for regular geometry mappings $\mathbf{F} \in \mathfrak{G}_D$. This means that any C^1-smooth isogeometric function is the push-forward of a glued spline function. This result is closely related to [9], which establishes the fact that matched G^k-constructions always yield C^k-continuous isogeometric elements in a more general setting.

Following the approach in [4], we construct a basis of the space $\mathscr{V}_F$—and consequently of the space $\mathfrak{G}_D$—by splitting the space into a direct sum of two subspaces, i.e.

$$\mathscr{V}_F = \mathscr{V}_F^\Gamma \oplus \mathscr{V}_F^S.$$

The first subspace $\mathscr{V}_F^\Gamma$ denotes the *space of interface functions*. It contains the functions with non-zero coefficients on the shared face Γ and the two neighboring layers, see blue points in Fig. 2. These functions are affected by the specific choice of gluing data concerned. The second subspace $\mathscr{V}_F^S$, referred to as the *space of standard functions*, contains functions with zero coefficients on these three layers, see the green bullets in Fig. 2. It is spanned by the "usual" isogeometric basis functions and therefore does not depend on the choice of gluing data.

In order to keep the presentation concise, we restrict ourselves to spaces satisfying first order homogeneous boundary conditions along $\partial \Omega$, which is indicated by the subscript 0, and we obtain the decomposition

$$\mathscr{V}_{F,0} = \mathscr{V}_{F,0}^\Gamma \oplus \mathscr{V}_{F,0}^S.$$

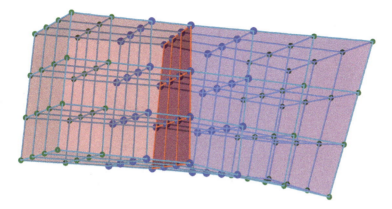

Fig. 2 Coefficients used by the interface functions (blue bullets) and coefficients contributing to inner functions (green bullets). The figure shows the control cages of both patches

3 Basis Construction

As noted in [4], trilinear geometric gluing data is particularly promising for applications. We describe the construction of a basis for $\mathscr{V}_{F,0}^{\Gamma}$ in this case.

This type of gluing data is derived from a trilinear geometry mapping $F = (F^{(1)}, F^{(2)})$, which is assumed to be regular. More precisely, it consists of the four polynomials

$$\beta(u,v) = \det\left(\partial_2 F^{(1)}(u,v,0), \partial_3 F^{(1)}(u,v,0), \partial_3 F^{(2)}(u,v,0)\right),$$

$$\gamma(u,v) = \det\left(\partial_1 F^{(1)}(u,v,0), \partial_3 F^{(1)}(u,v,0), \partial_3 F^{(2)}(u,v,0)\right),$$

$$\alpha^{(1)}(u,v) = \det \nabla F^{(2)}(u,v,0),$$

$$\alpha^{(2)}(u,v) = \det \nabla F^{(1)}(u,v,0), \qquad (2)$$

which have bi-degrees $[(3,2),(2,3),(2,2),(2,2)]$. For $p \geq 3$, the dimension of the space $\mathscr{V}_{F,0}^{\Gamma}$ is equal to

$$\dim \mathscr{V}_{F,0}^{\Gamma} = 10 + k(2 - 11k) - p(2 + 2k - 4k^2).$$

Locally supported basis functions for trilinear geometric gluing data were presented in [4]. For the sake of completeness, we recall the obtained coefficient patterns in Fig. 3. The basis for degree $p = 3$ is obtained by performing index shifts in $(2\mathbb{Z})^2$ for Type 3, while the basis for $p = 4$ consists of seven different types:

- $2k - 1$ functions of Type 4.1, with shifts in $2 \cdot 3\mathbb{Z}$,
- $k - 1$ functions of Type 4.2.1, with shifts in $3\mathbb{Z}$,

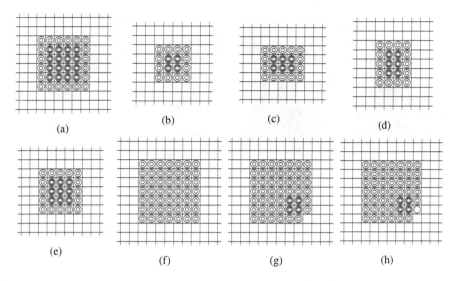

Fig. 3 Coefficient patterns of basis functions for trilinear geometric gluing data. (**a**) Type 3. (**b**) Type 4.1. (**c**) Type 4.2.1. (**d**) Type 4.2.2. (**e**) Type 4.3. (**f**) Type 4.4. (**g**) Type 4.5. (**h**) Type 4.6

- $k(k-1)$ functions of Type 4.2.2, with shifts in $(3\mathbb{Z})^2$, and
- $(k-1)^2$ functions of Types 4.3–4.6, with shifts in $(3\mathbb{Z})^2$.

The coefficient patterns of the basis functions described above were found by studying the kernel of the matrix formed by collocating the Eq. (1) at suitable Greville abscissa. The first order homogeneous boundary conditions are incorporated by imposing additional constraints. The corank of the resulting matrix reveals the dimension of the spline space $\mathcal{V}_{F,0}$, and consequently also of the subspace $\mathcal{V}_{F,0}^{\Gamma}$. Repeated patterns were observed in suitably constructed kernel vectors of that matrix, which allowed us to derive local subproblems that yield a single basis function.

To point out the importance of these local patterns, note that the computation of the sparsest kernel vectors (that is, the functions with the smallest possible support) is NP-hard. Therefore, even for small numbers of inner knots k, the computations can be rather inefficient. Another issue is that computing the rank and the corank of a matrix with floating point numbers can only be done up to certain precision. This can create some ambiguity or uncertainty on the actual dimension of the spline space.

In [4], the first problem was addressed by manually designing suitable orderings of the coefficients that lead to sparse coefficient patterns when performing RREF[1] computations. We dealt with the second issue by using rational arithmetic. In particular, the matrix obtained from (1) has rational elements, since they are

---

[1] Reduced Row Echelon Form.

Table 1 Time needed for constructing the basis using floating point and rational computations

	$k=3$	$k=7$	$k=15$	$k=31$
Floating point	0.01 s	0.07 s	1.03 s	25.82 s
Rational	5.73 s	1037.56 s	>24 h	>72 h

Table 2 Memory needed for constructing the bases using floating point and rational computations

	$k=3$	$k=7$	$k=15$	$k=31$
Floating point	9.9 MBytes	11.4 MBytes	17.6 MBytes	42.4 MBytes
Rational	20.8 MBytes	426.8 MBytes	>17.7 GBytes	>32 GBytes[2]

evaluations of piecewise polynomial functions with rational coefficients at rational Greville points. However, as the dimension of the problem increases, working with rational arithmetic becomes prohibitive.

In the present work, we exploit the fact that the local subproblems, which are defined by the shifted coefficient patterns, are known to have a kernel dimension equal to one. Consequently, it is no longer necessary to use rational arithmetic. Instead, since we know that we are looking for a single kernel vector, we use floating point computations and perform singular value decompositions. We then keep the vector associated to the singular value closest to zero.

The savings in time and memory needed for the basis computation when using floating point operations instead of exact arithmetic are presented in Tables 1 and 2. The entries of the tables refer to the construction of the basis of the space $\mathscr{V}_{F,0}^{\Gamma}$ for spline degree $p=3$ and different numbers of inner knots k. We compare the approach used in [4], where we had to set up the complete matrix obtained from (1) using rational arithmetic, with the more efficient construction using local subproblems and SVD based on floating point operations.

The entries in Table 1 show the time spent on the computations, whereas the values in Table 2 depict the maximal resident set size (RSS), which is the amount of memory occupied by the computation that is held in main memory (RAM). The expected massive advantage of the localized computation using floating point operations is clearly visible.

4 Approximation Properties

We explore the approximation power of the basis obtained for trilinear geometric gluing data in case of spline degree $p=3,4$. To determine the rates of convergence we use L^2-fitting of suitable target functions that are defined on the two-patch

---

[2] Aborted because of too high memory requirements.

Fig. 4 Two-patch geometry used for L^2-fitting

geometry shown in Fig. 4. Besides the usual norms, which are defined on the entire domain, we analyze the residuals via the following norms on the interface:

Type	$H^1(\Gamma)$	$L^2(\Gamma)$	$L^\infty(\Gamma)$
Norm	$\|\nabla(f^{(1)} - f^{(2)})|_\Gamma\|_2$	$\sqrt{\int_\Gamma |f|^2 ds}$	$\max_\Gamma |f|$

For bivariate C^1-smooth spline spaces, optimal convergence rates were obtained for degree $p = 3$ and higher, see [13]. However, this does not extend to the trivariate case of trilinearly parameterized two-patch domains, as shown in Fig. 5. In the left picture, the global error is shown, where a small reduction in the approximation power can be recognized. This loss in the convergence rate is solely introduced by the error on the interface, which is depicted in the right plot.

The reduction can be explained by taking a closer look on the involved spline functions. Consider two spline functions $f, \hat{f} \in \mathfrak{G}_D$, with $f|_\Gamma = \hat{f}|_\Gamma$. Since both functions are elements of the glued spline space $\mathfrak{G}_D$, they satisfy the following equation

$$\beta\, \partial_u g^{(1)} - \gamma\, \partial_v g^{(1)} + \alpha^{(1)}\, \partial_w g^{(1)} - \alpha^{(2)}\, \partial_w g^{(2)} = 0,$$

with $g \in \mathfrak{G}_D$, $g^{(1)} = g|_{\Omega^{(1)}}$, $g^{(2)} = g|_{\Omega^{(2)}}$, see (1).

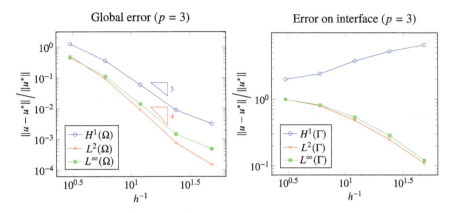

Fig. 5 L^2-approximation errors for trilinear geometric gluing data of degree $p = 3$

h	$H^1(\Omega)$	$L^2(\Omega)$	$L^\infty(\Omega)$	$H^1(\Gamma)$	$L^2(\Gamma)$	$L^\infty(\Gamma)$
0.33	1.29	0.52	0.46	2	1	1
0.17	0.36	$9.27 \cdot 10^{-2}$	0.11	2.42	0.8	0.82
$8.33 \cdot 10^{-2}$	$6.00 \cdot 10^{-2}$	$9.19 \cdot 10^{-3}$	$1.40 \cdot 10^{-2}$	3.76	0.48	0.54
$4.17 \cdot 10^{-2}$	$8.97 \cdot 10^{-3}$	$7.66 \cdot 10^{-4}$	$1.50 \cdot 10^{-3}$	5.24	0.25	0.29
$2.08 \cdot 10^{-2}$	$3.21 \cdot 10^{-3}$	$1.50 \cdot 10^{-4}$	$4.94 \cdot 10^{-4}$	6.57	0.11	0.12

Therefore, their difference, which is again a function in the glued spline space, satisfies

$$\beta \underbrace{\partial_u (f^{(1)} - \hat{f}^{(1)})|_\Gamma}_{=0} - \gamma \underbrace{\partial_v (f^{(1)} - \hat{f}^{(1)})|_\Gamma}_{=0} + \alpha^{(1)} \partial_w (f^{(1)} - \hat{f}^{(1)})|_\Gamma - \alpha^{(2)} \partial_w (f^{(2)} - \hat{f}^{(2)})|_\Gamma = 0,$$

hence

$$\alpha^{(1)} \partial_w (f^{(1)} - \hat{f}^{(1)})|_\Gamma = \alpha^{(2)} \partial_w (f^{(2)} - \hat{f}^{(2)})|_\Gamma.$$

It has been observed that $\gcd(\alpha^{(1)}, \alpha^{(2)}) = 1$, see [5]. This implies that $\alpha^{(1)}$, which is a bivariate polynomial of degree $(2,2)$, is a factor of each of the polynomial segments of

$$d^{(2)} = \partial_w (f^{(2)} - \hat{f}^{(2)})|_\Gamma.$$

Similarly, $\alpha^{(2)}$ is a factor of each of the polynomial segments of

$$d^{(1)} = \partial_w(f^{(1)} - \hat{f}^{(1)})|_\Gamma.$$

We obtain two C^1-smooth piecewise polynomial functions $d^{(1)}/\alpha^{(2)}$ and $d^{(2)}/\alpha^{(1)}$. Since the degree of these functions does not exceed $(3, 3) - (2, 2) = (1, 1)$, we conclude that the spline functions

$$d^{(i)} = \partial_w(f^{(i)} - \hat{f}^{(i)})|_\Gamma, \quad i = 1, 2$$

are indeed single polynomials of degree $(3, 3)$ and therefore C^∞-smooth.

Consequently, the cross-boundary derivatives of any two functions f and $\hat{f}$, which take the same values on the interface Γ, differ only by a bi-cubic polynomial with only four degrees of freedom, and this does not change as h is decreased. This observation, which is in agreement with the results (that were obtained by a slightly different approach) in [5], explains the loss of approximation power, as follows.

We consider two smooth functions $\varphi, \hat{\varphi} \in C^\infty(\Omega)$. There exist two sequences $(f_h)_h$ and $(\hat{f}_h)_h$ of trivariate tensor-product spline functions with uniform knots, whose elements are taken from the glued spline spaces obtained for element size

$$h = \frac{1}{k+1} \to 0,$$

such that $(f_h \circ F^{-1})_h$ and $(\hat{f}_h \circ F^{-1})_h$ converge to φ and $\hat{\varphi}$, respectively. If these sequences converged with the full approximation power, the derivatives would converge as well, hence

$$d_h^{(i)} = \partial_w(f_h^{(i)} - \hat{f}_h^{(i)})|_\Gamma \to \left(\partial_w((\varphi - \hat{\varphi}) \circ F^{(i)})\right)|_\Gamma, \quad i = 1, 2,$$

as $h \to 0$. However, this is impossible for almost all pairs of given functions $\varphi, \hat{\varphi}$, since the functions $d_h^{(i)}$ are single bicubic polynomials for any h.

If we consider spline degrees $p > 3$, this argument no longer applies and we observe full approximation power, as shown in Fig. 6 for $p = 4$.

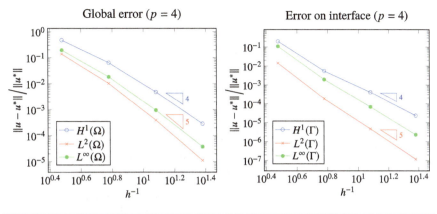

h	$H^1(\Omega)$	$L^2(\Omega)$	$L^\infty(\Omega)$	$H^1(\Gamma)$	$L^2(\Gamma)$	$L^\infty(\Gamma)$
0.33	0.47	0.14	0.19	0.2	$1.41 \cdot 10^{-2}$	0.11
0.17	$6.38 \cdot 10^{-2}$	$1.02 \cdot 10^{-2}$	$1.83 \cdot 10^{-2}$	$5.40 \cdot 10^{-3}$	$1.83 \cdot 10^{-4}$	$1.90 \cdot 10^{-3}$
$8.33 \cdot 10^{-2}$	$4.83 \cdot 10^{-3}$	$3.92 \cdot 10^{-4}$	$9.49 \cdot 10^{-4}$	$3.99 \cdot 10^{-4}$	$4.66 \cdot 10^{-6}$	$6.75 \cdot 10^{-5}$
$4.17 \cdot 10^{-2}$	$2.89 \cdot 10^{-4}$	$1.12 \cdot 10^{-5}$	$3.76 \cdot 10^{-5}$	$2.36 \cdot 10^{-5}$	$1.17 \cdot 10^{-7}$	$2.31 \cdot 10^{-6}$

Fig. 6 L^2-approximation errors for trilinear geometric gluing data of degree $p = 4$

5 Conclusion

Based on earlier results about locally supported bases on trivariate two-patch domains, which were shown to exist for trilinear geometric gluing data (that corresponds to piecewise trilinear domain parameterizations) and sufficiently high degree, we investigated the approximation properties of these functions. In addition we also showed how to efficiently compute the basis functions using standard arithmetic (i.e., floating point numbers). We observed that the existence of locally supported interface basis functions for spline degree $p = 3$ does not suffice to provide optimal approximation power, even though these functions take non-zero values along the interface. In addition to the experimental results we also derived a theoretical justification for this surprising fact. We also confirmed that these effects are no longer present for higher polynomial degrees. Future work will be devoted to multi-patch domains with more than two patches and to applications in numerical simulation.

Acknowledgments Supported by the Austrian Science Fund (FWF) through NFN S117 "Geometry + Simulation". The authors thank the reviewer for the constructive comments.

References

1. A. Bartezzaghi, L. Dedè, and A. Quarteroni. Isogeometric analysis of high order partial differential equations on surfaces. *Comput. Methods Appl. Mech. Engrg.*, 295:446–469, 2015.
2. L. Beirao da Veiga, A. Buffa, C. Lovadina, M. Martinelli, and G. Sangalli. An isogeometric method for the Reissner–Mindlin plate bending problem. *Comput. Methods Appl. Mech. Engrg.*, 209:45–53, 2012.
3. M. Bercovier and T. Matskewich. *Smooth Bézier Surfaces over Unstructured Quadrilateral Meshes*. Lecture Notes of the Unione Matematica Italiana, Springer, 2017.
4. K. Birner, B. Jüttler, and A. Mantzaflaris. Bases and dimensions of C^1-smooth isogeometric splines on volumetric two-patch domains. *Graph. Mod.*, 99:46–56, 2018.
5. K. Birner and M. Kapl. The space of C^1-smooth isogeometric spline functions on trilinearly parameterized volumetric two-patch domains. *Comput. Aided Geom. Des.*, 70:16–30, 2019.
6. A. Collin, G. Sangalli, and T. Takacs. Analysis-suitable G^1 multi-patch parametrizations for C^1 isogeometric spaces. *Comput. Aided Geom. Des.*, 47:93–113, 2016.
7. J. A. Cottrell, T. Hughes, and Y. Bazilevs. *Isogeometric Analysis: Toward Integration of CAD and FEA*. John Wiley & Sons, Chichester, England, 2009.
8. L. Dedè, M. J. Borden, and T. J. Hughes. Isogeometric analysis for topology optimization with a phase field model. *Arch. Comput. Methods Engrg.*, 19:427–465, 2012.
9. D. Groisser and J. Peters. Matched G^k-constructions always yield C^k-continuous isogeometric elements. *Comput. Aided Geom. Des.*, 34:67–72, 2015.
10. M. Kapl, F. Buchegger, M. Bercovier, and B. Jüttler. Isogeometric analysis with geometrically continuous functions on planar multi-patch geometries. *Comput. Methods Appl. Mech. Engrg.*, 316:209–234, 2017.
11. M. Kapl, G. Sangalli, and T. Takacs. Construction of analysis-suitable G^1 planar multi-patch parameterizations. *Comput. Aided Des.*, 97:41–55, 2018.
12. M. Kapl and V. Vitrih. Space of C^2-smooth geometrically continuous isogeometric functions on planar multi-patch geometries: Dimension and numerical experiments. *Comput. Math. Appl.*, 73(10):2319–2338, 2017.
13. M. Kapl, V. Vitrih, B. Jüttler, and K. Birner. Isogeometric analysis with geometrically continuous functions on two-patch geometries. *Comput. Math. Appl.*, 70(7):1518–1538, 2015.
14. B. Mourrain, R. Vidunas, and N. Villamizar. Dimension and bases for geometrically continuous splines on surfaces of arbitrary topology. *Comput. Aided Geom. Des.*, 45:108–133, 2016.
15. T. Nguyen, K. Karčiauskas, and J. Peters. C^1 finite elements on non-tensor-product 2d and 3d manifolds. *Appl. Math. and Comp.*, 272:148–158, 2016.
16. J. Peters. Geometric continuity. In *Handbook of computer aided geometric design*, pages 193–227. North-Holland, Amsterdam, 2002.

A Novel Approach to Fluid-Structure Interaction Simulations Involving Large Translation and Contact

Daniel Hilger, Norbert Hosters, Fabian Key, Stefanie Elgeti, and Marek Behr

Abstract In this work, we present a novel method for the mesh update in flow problems with moving boundaries, the phantom domain deformation mesh update method (PD-DMUM). The PD-DMUM is designed to avoid remeshing; even in the event of large, unidirectional displacements of boundaries. The method combines the concept of two mesh adaptation approaches: (1) The virtual ring shear-slip mesh update method (VR-SSMUM); and (2) the elastic mesh update method (EMUM). As in the VR-SSMUM, the PD-DMUM extends the fluid domain by a phantom domain; the PD-DMUM can thus locally adapt the element density. Combined with the EMUM, the PD-DMUM allows the consideration of arbitrary boundary movements. In this work, we apply the PD-DMUM in two test cases. Within the first test case, we validate the PD-DMUM in a 2D Poiseuille flow on a moving background mesh. Subsequently the fluid-structure interaction (FSI) problem serves as a proof of concept. Within the FSI problem, isogeometric analysis and NURBS-enhanced finite elements are employed to ensure an accurate description of the moving boundaries and a consistent coupling along the FSI boundary. Moreover, we stress the advantages of the novel method as compared to conventional mesh update approaches.

1 Introduction

Many flow phenomena in technical processes, e.g., flows in liquid storage tanks, valve and piston flows, and in general all fluid-structure interaction problems involve moving boundaries. The moving boundaries can cause topological changes of the fluid domain which are important to consider when solving the flow problem.

D. Hilger (✉) · N. Hosters · F. Key · S. Elgeti · M. Behr
Chair for Computational Analysis of Technical Systems RWTH Aachen University, Aachen, Germany
e-mail: hilger@cats.rwth-aachen.de; hosters@cats.rwth-aachen.de; key@cats.rwth-aachen.de; elgeti@cats.rwth-aachen.de; behr@cats.rwth-aachen.de

© Springer Nature Switzerland AG 2021
H. van Brummelen et al. (eds.), *Isogeometric Analysis and Applications 2018*,
Lecture Notes in Computational Science and Engineering 133,
https://doi.org/10.1007/978-3-030-49836-8_3

The changes of the fluid domain can be described either implicitly or explicitly [4]. In the implicit description–also called interface capturing–the boundary deformations are recorded on a fixed background mesh. This strategy has the advantage that complex topology changes, e.g., breaking waves, can be resolved easily. Yet, the treatment of discontinuities, conservation of mass, and the imposition of boundary conditions are still challenging. Examples of interface-capturing methods are the volume-of-fluid method [7] or the level-set method [13]. In the explicit description– called interface tracking–the domain deformations are described directly through the movements of its boundaries. The mesh is restricted to the fluid domain and conforms with its boundaries. This ensures an accurate approximation of the fluid interface and allows the imposition of boundary conditions along the moving boundary. However, every time the topology of the domain is changed, the mesh must be adapted accordingly.

The straightforward approach to incorporate the domain deformation is remeshing, but since remeshing is always connected to a projection of the solution between the old and the new mesh configurations, it should be avoided if possible [11]. As an alternative to remeshing, mesh update methods can be used, where the current mesh is adapted to the changes of the domain.

Mesh update methods can be categorized into two groups: (1) Methods in which the position of the mesh nodes are updated according to a predefined deformation rule, and (2) those where the mesh update is described by an additional set of equations [20]. In order to implement mesh update methods based on a predefined deformation rule, the changes of the fluid must be known in advance. One examples is the shear-slip mesh update method [3], where large relative rotational motions are considered by means of a connectivity update inside a small layer of the mesh.

In case that the motions of the boundaries are a priori unknown, the positions of the internal mesh nodes need to be computed according to the displacements of the moving boundaries. There exist different strategies to compute the updated positions of the internal mesh nodes, for example PDE based methods as the elastic mesh update method (EMUM) [11], the concept of radial basis functions [5], or spring based methods [2]. Furthermore, these methods have been enhanced by edge swapping and vertex smoothing operations to retain an adequate mesh quality even in the event of large displacements [21, 1]. Nevertheless, all of the above mentioned methods do not provide a satisfactory solution when it comes to boundary movements that result in strongly constricted or expanded parts of the initial mesh, as it happens for example in valve flows. This is because the existing mesh cells are either heavily squeezed or stretched. In this case, remeshing of the fluid domain becomes inevitable.

In order to avoid the need for remeshing, we propose a new mesh deformation method for large unidirectional mesh movements on boundary conforming meshes. Therefore, we combine the EMUM and the recently introduced virtual ring shear-slip mesh update method (VR-SSMUM) [12]. The basic idea is here to perform the mesh update by means of the EMUM, but allow additional mesh cells to enter or exit the fluid domain. Thus, the squeezing and the stretching of mesh cells is prevented by the possibility to increase or decrease the local number of finite elements (FE). The new method is employed in conjunction with the deforming-spatial-domain/stabilized space-time (DSD/SST) approach [18], which is used to solve the flow problem on the changing domain.

We conclude the introduction of the novel mesh update method by presenting the PD-DMUM in a test case that serves as a proof of concept. The test case describes a fluid-structure interaction (FSI) problem where isogeometric analysis [9] and NURBS-enhanced finite elements [16] are employed to obtain an accurate description of the deforming boundaries and a consistent coupling [8] across the FSI interface.

The structure of this paper is as follows: In Sect. 2, we provide the governing equations of the flow problems we want to consider in the scope of this work. Further, we briefly summarize the DSD/SST method and the EMUM. The concept and the implementation of the new mesh update method are explained in Sect. 3. In Sect. 4, the validation and testing of the mesh update method is discussed by means of two test cases.

2 Governing Equations of Fluid Dynamics

The proposed mesh update method is developed specifically for flow problems with boundary conforming meshes involving large unidirectional boundary movements. In this section, we present the governing equations of the flow problems examined within this work and further, we give a brief summary on the numerical methods employed to solve them.

2.1 Governing Equations of Fluid Dynamics

Consider an incompressible fluid covering the deformable fluid domain $\Omega_t^f \subset R^{n_{sd}}$, with n_{sd} indicating the number of spatial dimensions. At every time instant $t \in [0, T]$, the fluid's unknown velocity $\mathbf{u}(\mathbf{x}, t)$ and pressure $p(\mathbf{x}, t)$ are governed by the Navier-Stokes equations for incompressible fluids:

$$\rho^f \left(\frac{\partial \mathbf{u}^f}{\partial t} + \mathbf{u}^f \cdot \nabla \mathbf{u}^f - \mathbf{f}^f \right) - \nabla \cdot \boldsymbol{\sigma}^f = \mathbf{0} \quad \text{on } \Omega_t^f, \forall t \in (0, T), \quad (1a)$$

$$\nabla \cdot \mathbf{u}^f = 0 \quad \text{on } \Omega_t^f, \forall t \in (0, T), \quad (1b)$$

with ρ^f denoting the fluid density and $\mathbf{f}^f$ representing all external body forces per unit mass. For Newtonian fluids, the stress tensor $\boldsymbol{\sigma}^f$ is defined as

$$\boldsymbol{\sigma}^f = -p^f \mathbf{I} + 2\rho^f \nu^f \boldsymbol{\varepsilon}^f(\mathbf{u}^f), \quad (2)$$

with

$$\varepsilon^f(\mathbf{u}^f) = \frac{1}{2}\left(\nabla \mathbf{u}^f + \left(\nabla \mathbf{u}^f\right)^T\right), \tag{3}$$

where ν^f denotes the dynamic viscosity. A well-posed system is obtained when boundary conditions are imposed on the external boundary Γ_t^f. Here, we distinguish between Dirichlet and Neumann boundary conditions given by:

$$\mathbf{u}^f = \mathbf{g}^f \quad \text{on } \Gamma_{t,g}^f, \tag{4a}$$

$$\mathbf{n}^f \cdot \boldsymbol{\sigma}^f = \mathbf{h}^f \quad \text{on } \Gamma_{t,h}^f, \tag{4b}$$

where $\mathbf{g}^f$ and $\mathbf{h}^f$ prescribe the velocity and stress values on complementary subsets of Γ_t^f. With regard to deformation of the fluid domain Ω_t^f in time, the DSD/SST method is applied to solve the Navier-Stokes equations.

2.2 Deforming-Spatial-Domain/Stabilized Space-Time Method

The DSD/SST method is a space-time-based finite-element (FE) method, i.e., a FE discretization is applied to space and time. It was first applied to flow problems with moving boundaries in [18, 19].

The advantage of the DSD/SST method is, that the variational form of the governing equations implicitly incorporates the deformations of the domain. In order to construct the interpolation and weighting function spaces used in the variational formulation of the problem, the time interval $(0, T)$ is split into N subintervals $I_n = [t_n, t_{n+1}]$, where t_n and t_{n+1} belong to an ordered series of time levels. Thus, the space-time continuum is divided into N space-time slabs Q_n as

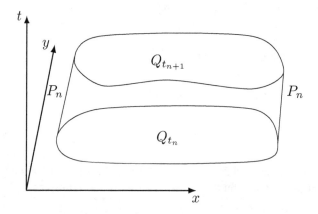

Fig. 1 Space-time slab

depicted in Fig. 1, bounded by the spatial configurations Ω_t at time t_n and t_{n+1}, and P_n describing the course of the spatial boundary Γ_t^f as t traverses I_n. The boundary P_n can be decomposed into two complementary subsets $(P_n)_g$ and $(P_n)_h$, representing the Dirichlet and Neumann boundary conditions of Γ_t^f $\forall t \in I_n$. The space-time slabs are weakly coupled along their interfaces using jump terms. For the spatial approximation $\Omega_{t,h}^f$ of the domain Ω_t^f, the following finite element trial and weighting function spaces are constructed:

$$\mathcal{H}^{1h}(Q_n) := \left\{ w^h \in \mathcal{H}^1(Q_n) \,\middle|\, w_T^h \text{ is a first-order polynomial } \forall T \in \mathcal{T}^h \right\}, \tag{5a}$$

$$\mathcal{S}_u^h := \left\{ \mathbf{u}^h | \mathbf{u}^h \in \left[\mathcal{H}^{1h}(Q_n)\right]^{nsd}, \mathbf{u}^h = \mathbf{g} \text{ on } (P_n)_g \right\}, \tag{5b}$$

$$\mathcal{V}^h := \left\{ \mathbf{w}^h | \mathbf{w}^h \in \left[\mathcal{H}^{1h}(Q_n)\right]^{nsd}, \mathbf{w}^h = \mathbf{0} \text{ on } (P_n)_g \right\}, \tag{5c}$$

$$\mathcal{S}_p^h = \mathcal{V}_p^h := \left\{ q^h | q^h \in \mathcal{H}^{1h}(Q_n) \right\}. \tag{5d}$$

Therein, the space $\mathcal{H}^1(Q_n)$ is approximated by the interpolation function space $\mathcal{H}^{1h}(Q_n)$ with first-order polynomial functions in space and time. Using the following notational convention,

$$\left(\mathbf{u}^h\right)_n^{\pm} = \lim_{\epsilon \to 0} \mathbf{u}(t_n \pm \epsilon) \tag{6a}$$

$$\int_{Q_n} \cdots dQ = \int_{I_n} \int_{\Omega_t} \cdots d\Omega dt, \tag{6b}$$

$$\int_{(P_n)} \cdots dP = \int_{I_n} \int_{\Gamma_t} \cdots d\Gamma dt, \tag{6c}$$

and following references [18, 10, 14], the stabilized variational formulation of the Navier Stokes equations is obtained: Given $\left(\mathbf{u}^h\right)_n^{-}$ with $\left(\mathbf{u}^h\right)_0^{-} = \mathbf{u}_0$, find $\mathbf{u}^h \in \mathcal{S}_{\mathbf{u}}^h$ and $p^h \in \mathcal{S}_p^h$ such that $\forall \mathbf{w}^h \in \mathcal{V}_{\mathbf{u}}^h, \forall q \in \mathcal{V}_p^h$:

$$\int_{Q_n} \mathbf{w}^h \cdot \rho^f \left(\frac{\partial \mathbf{u}^h}{\partial t} + \mathbf{u} \cdot \nabla \cdot \mathbf{u}^h - \mathbf{f} \right) dQ + \int_{Q_n} \nabla \mathbf{w}^h : \sigma(p^h, \mathbf{u}^h) dQ$$

$$+ \int_{Q_n} q^h \nabla \cdot \mathbf{u}^h dQ + \int_{\Omega_n} \left(\mathbf{w}^h\right)_n^{+} \cdot \rho^f \left(\left(\mathbf{u}^h\right)_n^{+} - \left(\mathbf{u}^h\right)_n^{-} \right) d\Omega$$

$$+ \sum_{e=1}^{n_{el}} \int_{Q_n^e} \frac{1}{\rho^f} \tau_{MOM} \left[\rho^f \mathbf{u}^h \cdot \nabla \mathbf{w}^h + \nabla q^h \right]$$

$$\cdot \left[\rho^f \left(\frac{\partial \mathbf{u}^h}{\partial t} + \mathbf{u} \cdot \nabla \cdot \mathbf{u}^h - \mathbf{f} \right) - \nabla \cdot \boldsymbol{\sigma}(p^h, \mathbf{u}^h) \right] d\Omega$$

$$+ \sum_{e=1}^{n_{el}} \int_{Q_n^e} \nabla \cdot \mathbf{w}^h \rho^f \tau_{CONT} \nabla \cdot \mathbf{u}^h d\Omega$$

$$= \int_{(P_n)_h} \mathbf{w}^h \cdot \mathbf{h}^h dP. \tag{7}$$

In Eq. (7), the first three terms and the last term directly result from the variational formulation of Eq. (1), whereas the fourth term denotes the jump terms between the space-time slabs. Terms five and six result from a Galerkin-Least Squares (GLS) stabilization applied to the Navier-Stokes equations. The stabilization approach used within this work and the choice of the stabilization parameters τ_{CONT} and τ_{MOM} are described in detail in [14].

Though the DSD/SST method implicitly accounts for the domain deformations in one time slab, a deformation rule is needed to deform the FE mesh according to the boundary movements.

2.3 Elastic Mesh Update Method

One approach for the automatic mesh update in boundary conforming meshes is the elastic mesh update method (EMUM) introduced by Johnson and Tezduyar [11], where the mesh is understood as an elastic body occupying the bounded region $\Omega^{\#} \subset \mathcal{R}^{n_{sd}}$ with boundary $\Gamma^{\#}$. Thus, the deformation of the mesh is expressed in terms of the nodal displacements $\mathbf{d}^{\#}$ governed by the equilibrium equation of elasticity:

$$\nabla \cdot \boldsymbol{\sigma}^{\#} = \mathbf{0}, \tag{8}$$

where $\boldsymbol{\sigma}^{\#}$ corresponds to the Cauchy stress tensor,

$$\boldsymbol{\sigma}^{\#} = \lambda \left(tr \boldsymbol{\epsilon}^{\#} \right) \mathbf{I} + 2\mu \boldsymbol{\epsilon}^{\#}, \qquad \boldsymbol{\epsilon}^{\#} = \frac{1}{2} \left(\nabla \mathbf{d}^{\#} + \left(\nabla \mathbf{d}^{\#} \right)^T \right). \tag{9}$$

The imposition of Dirichlet and Neumann boundary conditions yields a well-posed problem for the mesh deformation:

$$\mathbf{d}^{\#} = \mathbf{g}^{\#} \qquad \text{on } (\Gamma)_g^{\#}, \tag{10}$$

$$\mathbf{n} \cdot \boldsymbol{\sigma}^{\#} = \mathbf{h}^{\#} \qquad \text{on } (\Gamma)_h^{\#}, \tag{11}$$

where $\mathbf{g}^{\#}$ and $\mathbf{h}^{\#}$ prescribe the displacements and normal stresses on the mesh boundaries.

The elasticity problem is solved with the Galerkin FE method and the resulting displacements are applied to the mesh nodes representing the upper mesh configuration of the current space-time slab.

3 The Phantom Domain Mesh Deformation Method

The aim of the newly proposed method is to extend the usability of boundary-conforming meshes for deforming domains with large, unidirectional deformations. The specific target are applications with large, unidirectional deformations (imagine an object sinking within a fluid or the flow through a valve). So far, the fluid domain is enclosed within two types of boundaries: (1) deforming, and (2) fixed. The deforming boundaries are handled in a standard interface tracking way, meaning that the boundary deforms according to its relevant deformation rule—e.g., determined by the structure in an FSI context or a free-surface motion—while the inner nodes adapt to this motion. As depicted in Fig. 2, a predominantly unidirectional deformation, however, soon results in a situation where one side of the mesh contains very compressed elements, whereas the other side is comprised of very stretched elements.

In our proposed method, this is counteracted via the implementation of a new boundary condition that allows mesh cells to exit and enter the fluid domain as needed. The implementation of this boundary condition is based on the concept of the VR-SSMUM presented in [12]. As with the VR-SSMUM, the mesh is extended

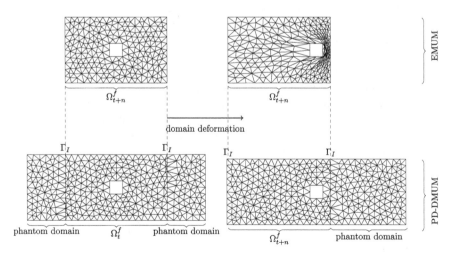

Fig. 2 Mesh deformation with PD-DMUM vs. EMUM

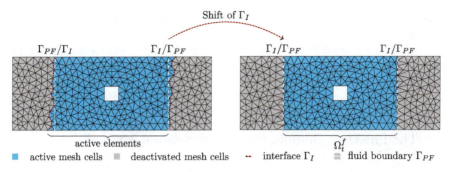

Fig. 3 Activity pattern on mesh with initial and corrected course of interface Γ_I

by additional mesh cells. The VR-SSMUM, however, is restricted to uniform mesh movements on a subset of the mesh and does not consider arbitrary boundary movements as needed for example in FSI computations. As sketched in Fig. 2, the additional cells are positioned in a phantom domain which is located outside of the fluid domain. In the following we will therefore refer to this method as the phantom domain deformation mesh update method (PD-DMUM).

Since not all mesh cells are positioned within the fluid domain, an activity pattern, as illustrated in Fig. 3, is used to determine which elements are used in the computation of the flow problem. Here, elements that intersect with the fluid domain are considered as activated elements whereas the remaining elements are deactivated. Activated and deactivated elements have a common interface Γ_I. The interface is a boundary of the fluid domain, which requires the definition of boundary values. The boundary value prescribed at the element nodes of the interface is of a new boundary type. The element nodes associated with the new boundary type have the special characteristic that they prescribe boundary values to the flow problem, but function as internal nodes in the mesh update method. Consequently, the mesh of the phantom domain and the fluid domain are considered as one coherent mesh in the mesh deformation process.

Now that the mesh is deformed according to the underlying deformation rule, elements from the phantom domain can slide across the prescribed fluid boundary Γ_{PF} into the fluid domain or vice versa. This changes the composition of elements that intersect with the fluid domain, so that the activity pattern of the elements must be re-determined. In the space-time approach used here, one space-time slab is bounded by two different mesh configurations. This can lead to the situation shown in Fig. 4a, where an element is located inside the fluid domain on the upper time level, yet positioned outside at the lower time level. Therefore, we define here that the mesh configuration at the upper time level always determines which elements represent the fluid domain. Based on the updated activity pattern, the new location of the interface Γ_I is determined within the mesh. The position of Γ_I usually does not correspond to the position of the predefined fluid boundary Γ_{PF}.

The boundary conformity of the mesh for Γ_{PF} is now obtained by a closest point projection of all mesh nodes on Γ_I to the prescribed contour of the fluid boundary.

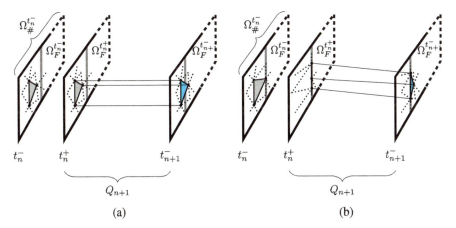

Fig. 4 Shifting of Γ_I for boundary conformity in space-time slab. (**a**) Uncorrected boundary Γ_I. (**b**) Corrected boundary Γ_I

Since the projection of mesh nodes can result in arbitrarily small elements, a tolerance which depends on the element size is considered in the determination process of the nodal activity. Further, it is important to note the special case of those elements which were not yet part of the fluid discretization in the previous time step, because these elements require a projection of the old solution onto the new boundary nodes. This is necessary to calculate the jump terms in Eq. (1). This means that the new method does not require remeshing, yet the projection between two mesh configurations cannot be completely avoided. However, the projection is limited to single elements when they enter the fluid domain.

The sequence of the individual steps within the PD-DMUM can be summarized as follows:

1. Update mesh according to moving boundaries.
2. **Identify activated and deactivated elements.**
3. **Adapt the boundaries to the prescribed position of the fluid domain.**
4. **Set boundary values for the nodes on the redefined interface Γ_I.**
5. **Project the solution of the previous time step for all newly activated elements.**
6. Solve flow problem on active elements.

In direct comparison with a conventional update strategy for boundary conforming meshes, such as the EMUM, steps (2)–(5) are those which are additionally required.

Depending on the boundary movements, the PD-DMUM can be complemented with additional mesh update strategies. In case of large unidirectional boundary movements, we can employ the concept of the virtual ring presented in [12]. The objective of the virtual ring is to reduce the size of the phantom domain in the mesh update. For this purpose, we connect the mesh along the outward facing boundaries of two oppositely positioned phantom domains. This connection results

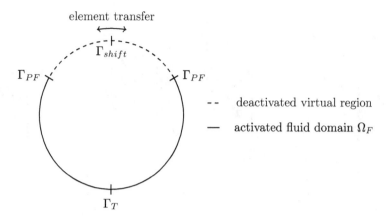

Fig. 5 Illustration of the virtual ring concept

in a coherent mesh, forming a virtual ring as illustrated in Fig. 5. The mesh update can now transfer elements between the connected phantom domains, while moving them along the virtual ring. Consequentially, elements can exit the fluid domain on one side and re-enter the domain on the other side. Therefore, the phantom domains can be reduced to a thin layer of elements. The reduction of the phantom domains results in a significant decrease of computational cost for the mesh deformation problem.

4 Computational Results

The implementation of the PD-DMUM is applied to two test cases. In a first step, we validate the mesh update method by examining its influence on the solution of a two dimensional Poiseuille flow. In the second test case we show, by means of an example from the field of FSI, the advantages of the PD-DMUM.

4.1 2D Poiseuille Flow on Moving Background Mesh

In the first test case we examine the influence of the PD-DMUM on a flow problem with a well-known solution. For this purpose, we consider a two-dimensional Poiseuille flow in a tube. The topology of the fluid domain remains unchanged, yet a predefined motion is applied to the underlying mesh. The PD-DMUM is used to perform the mesh update, but should not affect the flow field within the tube.

The geometric dimensions of the tube are chosen according to Fig. 6. In the middle of the domain, we position a mesh section Γ_T by means of which the

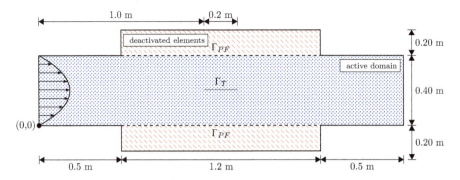

Fig. 6 Tube geometry for Poiseuille flow

Table 1 Properties of fluid in 2D Poiseuille flow

Parameter	Identifier	Value
Density	ρ	1.0 [kg/m³]
Viscosity	ν	0.001 [kg/m·s]
Mean velocity	U	2.5 [m/s]

predefined mesh motion is imposed as a Dirichlet boundary condition. The boundary Γ_T has no physical impact with respect to the flow problem. The additional phantom domains required within the PD-DMUM are positioned along the upper and lower boundary of the tube. The material properties of the fluid are chosen according to Table 1. Regarding boundary conditions of the flow, we impose no-slip condition along the walls of the tube. This also applies to the boundary section Γ_{PF} at the interface between the phantom domain and the fluid domain. A parabolic inflow profile for the velocity is given at the inlet of the tube:

$$\mathbf{u}(y) = \left(\frac{4U y(H-y)}{H^2}, 0 \right). \tag{12}$$

At the outflow the tangential velocity components are set to zero and the pressure is assumed to be $p = 0$. With respect to the mesh update, the position of the nodes at the inlet, the outlet, and the tube walls are fixed. However, this does not apply to Γ_{PF} and the remaining boundaries of the phantom domain, as these nodes should be able to move freely. For the boundary Γ_T we prescribe the following sinusoidal movement:

$$\mathbf{d}(t) = \left(0, \, 0.1 \cdot \sin\left(\frac{2\pi t}{T} \right) \right). \tag{13}$$

The mesh deformation is examined for a period of $T = 8$ [s]. The time step size is $\Delta t = 0.02$ [s]. Initially, a fully developed flow profile is already present in the pipe.

The Poiseuille flow is computed on four mesh configurations with the PD-DMUM and for the purpose of comparison for one configuration by the EMUM. For the comparison of the solutions we use the flow velocity. The velocity is measured at a probe positioned at point (1.1, 0.2) inside the tube. Together with the given analytical solution of the Poiseille flow, the relative error can be computed for the different mesh configurations.

In a first step, the relative error of the computed velocity is evaluated for the probe position. In Fig. 7 it can be observed that the relative error decreases as the mesh is refined. The comparison between the solution of the EMUM and the PD-DMUM on similar grids shows that the relative error for the calculated velocity is of the same order of magnitude. The fluctuations that can be observed for all computations can be explained by the linear interpolation of the parabolic velocity profile at the probe position. In Fig. 8, we can observe that the numerical solution converges for the PD-DMUM towards the analytic solution of the Poiseuille problem. Both, the convergence of the PD-DMUM and the comparable results to the EMUM for

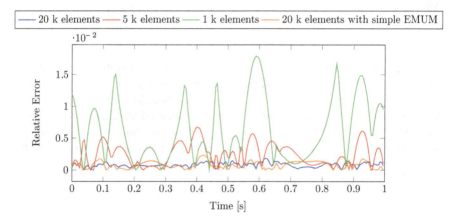

Fig. 7 Relative error of velocity at probe position

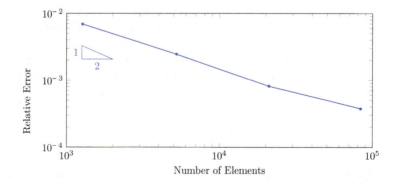

Fig. 8 Average relative error for different mesh resolutions

moderate mesh deformations indicate that the PD-DMUM provides a valid mesh update.

4.2 Falling Ring in a Fluid-Filled Container

The second test case is used to illustrate possible applications of the PD-DMUM. For this purpose, we consider a fluid-structure interaction with large translational boundary movement. More precisely, we simulate an elastic ring that falls inside a fluid-filled container until it hits the ground and rebounds. Concerning the mesh deformation, this is a demanding process, since the number of mesh cells, which are initially positioned between the ring and the bottom, must be reduced to zero by the time of contact. Using previous mesh update methods it is not possible to simulate this process on boundary conforming meshes without frequent remeshing of the fluid domain.

The geometric dimensions of the container and the ring are chosen according to Fig. 9. The ring is represented by a non-uniform rational B-spline (NURBS) [15] with 721 elements and second-order basis functions. In total 13,448 elements are used to discretize the fluid domain and the additional phantom domains. In the flow problem no-slip conditions are prescribed along the walls and the bottom of the container, whereas the top of the container is assumed to be open. The fluid velocity at the ring surface corresponds to the structural velocity. In terms of the mesh deformation problem the mesh nodes on the container and walls of the

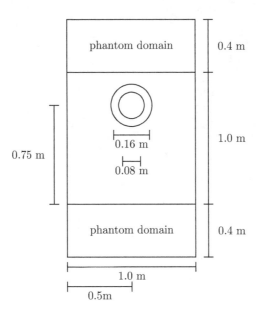

Fig. 9 Geometry of container with ring

phantom domains are restricted to a vertical movement. The structural deformation is prescribed as a Dirichlet value for the ring boundary.

The FSI problem is solved in a staggered solution approach with an implicit coupling scheme [6]: The fluid loads and the structural displacements are exchanged iteratively until both field problems are fully converged. On the structural side, the deformation of the ring are represented by a linear elastic problem solved with isogeometric analysis (IGA) [9]. The contact interaction between the ring and the bottom of the container is considered via the penalty method [17]. The flow field induced by the motion of the ring is described by the Navier-Stokes equations which are solved by the DSD/SST approach in combination with the presented PD-DMUM. The two field problems are strongly coupled in time [20], and for the spatial coupling we apply a NURBS-based coupling following [8].

In Figs. 10, 11, 12, 13, and 14, we present snapshots of the simulation at different points in time, starting from the initial position of the ring, via the moment when the ring is in contact with the bottom of the container, up to the point of maximal altitude after the first contact interaction. As it can be guessed from the snapshot in Fig. 12, one element remains between the bottom of the container and the falling ring. This element will not be removed because we cannot exactly comply with the contact conditions using the penalty method. Nevertheless, it can be observed in every snapshot, that mesh cells experience large displacements but only little deformations. Due to the application of the PD-DMUM, the entire FSI problem was solved without remeshing.

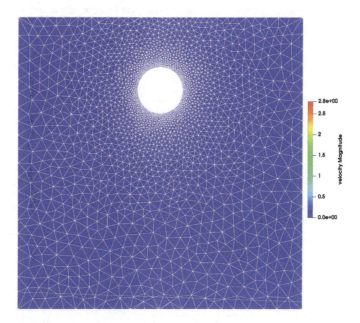

Fig. 10 Velocity at $t = 0$ s

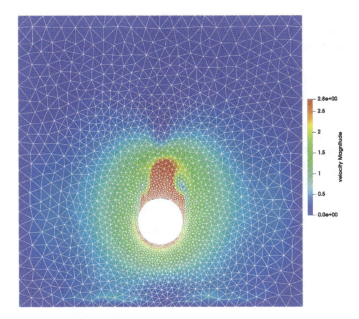

Fig. 11 Velocity at $t = 0.55$ s

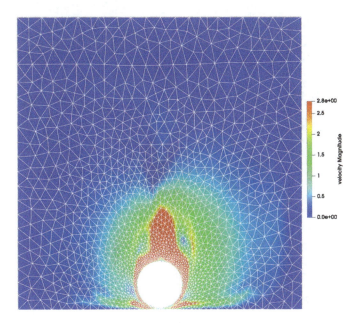

Fig. 12 Velocity at $t = 0.75$ s

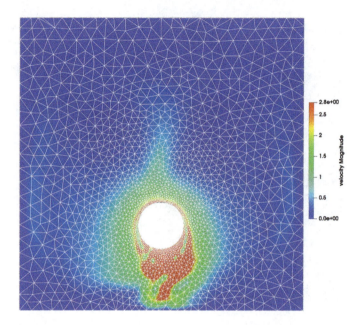

Fig. 13 Velocity at $t = 1.0$ s

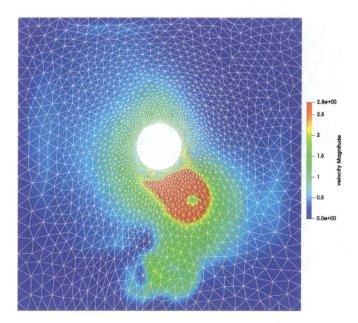

Fig. 14 Velocity at $t = 1.45$ s

5 Discussion

In this paper, we presented a novel approach for the mesh update of boundary conforming meshes, particularly developed for problems with large unidirectional boundary movements, the PD-DMUM. Subsequent to the description, we evaluated the PD-DMUM in two test cases. In the first test case we showed by means of a Poiseuille flow the general agreement of the PD-DMUM with results of consisting methods. In the fluid-structure interaction problem presented in the second test case we emphasised the applicability of the PD-DMUM in complex processes with moving boundaries.

Acknowledgments This work was supported by the German Research Foundation under the Cluster of Excellence "Integrative production technology for high-wage countries" (EXC128) as well as the German Research Foundation under the Cluster of Excellence "Internet of Production". Computing resources were provided by the AICES graduate school and RWTH Aachen University Center for Computing.

References

1. F. Alauzet: Efficient moving mesh technique using generalized swapping. Proceedings of the 21st International Meshing Roundtable. Springer, Berlin, Heidelberg, 17–37 (2003).
2. J.T. Batina: Unsteady Euler airfoil solutions using unstructured dynamic meshes. AIAA journal, vol. **28**, 1381–1388 (1990).
3. M. Behr, and D. Arora: Shear-slip mesh update method: Implementation and applications. Computer Methods in Biomechanics and Biomedical Engineering, vol. **6**, 113–123, (2003).
4. S. Elgeti, and H. Sauerland: Deforming fluid domains within the finite element method: five mesh-based tracking methods in comparison. Archives of Computational Methods in Engineering, vol. **23**, 323–361, (2016).
5. A. De Boer, M. S. Van der Schoot, and H. Bijl: Mesh deformation based on radial basis function interpolation. Computers & structures **85** 784–795 (2007).
6. C.A. Felippa, K.C. Park, and C. Farhat: Partitioned analysis of coupled mechanical systems. Computer methods in applied mechanics and engineering, vol. **190**, 3247–3270 (2001).
7. C.W. Hirt, and B.D. Nichols: Volume of fluid (VOF) method for the dynamics of free boundaries. Journal of computational physics, vol. **39** 201–225 (1981).
8. N. Hosters, J. Helmig, A. Stavrev, M. Behr, and S. Elgeti: Fluid-Structure Interaction with NURBS-Based Coupling. Computer Methods in Applied Mechanics and Engineering, vol. **332** 520–539 (2018).
9. T.J.R. Hughes, J.A. Cottrell, and Y. Bazileves. Isogeometric analysis: CAD, finite elements, NURBS, exact geometry and mesh refinement. Computer Methods in Applied Mechanics and Engineering, vol. **194**, 4135–4195 (2005).
10. T.J.R. Hughes, L.P. Franca, and G.M. Hulbert: A new finite element formulation for computational fluid dynamics: VIII. The Galerkin/least-squares method for advective-diffusive equations. Computer Methods in Applied Mechanics and Engineering, vol. **73**, 173–189 (1989).
11. A.A. Johnson, and T.E. Tezduyar: Mesh update strategies in parallel finite element computations of flow problems with moving boundaries and interfaces. Computational Methods in Applied Mechanical Engineering, vol. **119** 73–94 (1994).

12. F. Key, L. Pauli, and S. Elgeti: The Virtual Ring Shear-Slip Mesh Update Method. Computer and Fluids, vol. **172**, 352–361 (2018).
13. S. Osher, and J.A. Sethian: Fronts propagating with curvature-dependent speed: algorithms based on Hamilton-Jacobi formulations. Journal of computational physics, vol. **79** 12–49 (1988).
14. L. Pauli, and M. Behr: On stabilized space-time FEM for anisotropic meshes: Incompressible Navier–Stokes equations and applications to blood flow in medical devices. International Journal for Numerical Methods in Fluids, vol. **85**, 189–209 (2017).
15. L. Piegl, and W. Tiller: The NURBS book. Springer, Berlin, (1997).
16. R. Sevilla, S. Fernández-Méndez and A. Huerta: NURBS-enhanced finite element method (NEFEM). International Journal for Numerical Methods in Engineering **76.1** 56–83 International Journal for Numerical Methods in Engineering 76.1 (2008): 56–83.
17. I. Temizer, P. Wriggers, and T.J.R. Hughes: Contact Treatment in Isogeometric Analysis with NURBS". Computer Methods in Applied Mechanics and Engineering vol. **200**, 1100–1112 (2011).
18. T.E. Tezduyar, M. Behr, and J. Liou: A new strategy for finite element computations involving moving boundaries and interfaces – the deforming-spatial-domain/space-time procedure: I The concept and the preliminary numerical tests. Computational Methods in Applied Mechanical Engineering, vol. **94**, 339–351 (1992).
19. T.E. Tezduyar, M. Behr, S. Mittal, and J. Liou: A new strategy for finite element computations involving moving boundaries and interfaces – the deforming-spatial-domain/space-time procedure: II Computations of free-surface flows, two-liquid flows, and flows with drifting cylinders. Computational Methods in Applied Mechanical Engineering, vol. **94**, 353–371 (1992).
20. W.A. Wall: Fluid-Struktur-Interaktionen mit stabilisierten Finiten Elementen". Institut für Baustatistik der Universität Stuttgart, (1999).
21. L. Wang, and PO. Persson: A high-order discontinuous Galerkin method with unstructured space-time meshes for two-dimensional compressible flows on domains with large deformations. Computers & Fluids, vol. 118, 53–68 (2015).

An IGA Framework for PDE-Based Planar Parameterization on Convex Multipatch Domains

Jochen Hinz, Matthias Möller, and Cornelis Vuik

Abstract The first step towards applying isogeometric analysis techniques to solve PDE problems on a given domain consists in generating an analysis-suitable mapping operator between parametric and physical domains with one or several patches from no more than a description of the boundary contours of the physical domain. A subclass of the multitude of the available parameterization algorithms are those based on the principles of *Elliptic Grid Generation* (EGG) which, in their most basic form, attempt to approximate a mapping operator whose inverse is composed of harmonic functions. The main challenge lies in finding a formulation of the problem that is suitable for a computational approach and a common strategy is to approximate the mapping operator by means of solving a PDE-problem. PDE-based EGG is well-established in classical meshing and first generalization attempts to spline-based descriptions (as is mandatory in IgA) have been made. Unfortunately, all of the practically viable PDE-based approaches impose certain requirements on the employed spline-basis, in particular global $C^{\geq 1}$-continuity.

This paper discusses an EGG-algorithm for the generation of planar parameterizations with locally reduced smoothness (i.e., with support for locally only C^0-continuous bases). A major use case of the proposed algorithm is that of multipatch parameterizations, made possible by the support of C^0-continuities. This paper proposes a specially-taylored solution algorithm that exploits many characteristics of the PDE-problem and is suitable for large-scale applications. It is discussed for the single-patch case before generalizing its concepts to multipatch settings. This paper is concluded with three numerical experiments and a discussion of the results.

J. Hinz (✉) · M. Möller · C. Vuik
Delft Institute of Applied Mathematics, Delft, The Netherlands
e-mail: jochen.hinz@epfl.ch; m.moller@tudelft.nl; c.vuik@tudelft.nl

© Springer Nature Switzerland AG 2021
H. van Brummelen et al. (eds.), *Isogeometric Analysis and Applications 2018*,
Lecture Notes in Computational Science and Engineering 133,
https://doi.org/10.1007/978-3-030-49836-8_4

1 Introduction

The automatic generation of analysis-suitable planar parameterizations for IgA-based numerical simulations is a difficult, yet important problem in the field of isogeometric analysis, since generally no more than a description of the boundary contours is available. The main challenge lies in the generation of a folding-free (i.e., bijective) parameterization with numerically favorable properties such as orthogonal isolines and a large degree of parametric *smoothness*. Furthermore, a practical algorithm should be computationally inexpensive, and, if possible, exhibit little sensitivity to small perturbations in the boundary contour description.

Let Ω denote the target geometry and $\hat{\Omega}$ the parametric domain. Furthermore, let $\mathbf{x} : \hat{\Omega} \to \Omega$ denote the mapping operator that we attempt to build from the linear span of the B-Spline basis $\Sigma = \{w_1, w_2, \ldots, w_N\}$, where $\mathbf{x}|_{\partial\hat{\Omega}} = \partial\Omega$ is known. Note that $\mathbf{x}$ is of the form:

$$\mathbf{x}(\xi, \eta) = \sum_{i \in \mathcal{I}_{\text{boundary}}} \mathbf{c}_i w_i(\xi, \eta) + \sum_{j \in \mathcal{I}_{\text{inner}}} \mathbf{c}_j w_j(\xi, \eta), \qquad (1)$$

where $\mathcal{I}_{\text{inner}}$ and $\mathcal{I}_{\text{boundary}}$ denote the index set of the vanishing and nonvanishing basis functions on $\partial\hat{\Omega}$, respectively. Formally, $\mathcal{I}_{\text{boundary}} \cap \mathcal{I}_{\text{inner}} = \emptyset$ and $\mathcal{I}_{\text{boundary}} \cup \mathcal{I}_{\text{inner}} = \{1, \ldots, N\}$. With this, the objective of all parameterization algorithms is to properly select the inner control points $\mathbf{c}_j$, while the boundary control points $\mathbf{c}_i$ are known from the boundary contours and typically held fixed.

In [8], Gravesen et al. study planar parameterization techniques based on the constrained minimization of a quality functional over the inner control points. To avoid self-intersections, a nonlinear and nonconvex sufficient condition for $\det J > 0$, where J denotes the Jacobian of the mapping, is added as a constraint. The numerical quality of the resulting parameterization depends on the choice of the employed cost functional and the characteristic properties of Ω. While this approach is not guaranteed to yield acceptable results for all types of geometries (see Sect. 4), it is known to yield good results in a wide range of applications with proper parameter tuning. A drawback is the relatively large number of required iterations (typically ~ 30) and the need to find an initial guess that satisfies the constraints (for which another optimization problem has to be solved first). The proposed minimization is tackled with a black-box nonlinear optimizer (IPOPT [2]) that comes with all the drawbacks of nonlinear optimization such as the danger of getting stuck in local minima.

Another class of parameterization methods suitable for nontrivial geometries are PDE-based, most notably, the class of methods based on the principles of *elliptic grid generation* (EGG). Methods based on EGG attempt to generate a mapping $\mathbf{x} : \hat{\Omega} \to \Omega$ such that the components of $\mathbf{x}^{-1} : \Omega \to \hat{\Omega}$ are harmonic functions on Ω. For this, a nonlinear partial differential equation is imposed on $\mathbf{x}$, which takes the form

$$\mathcal{L}(\mathbf{x}) = g_{22}\mathbf{x}_{\xi\xi} - 2g_{12}\mathbf{x}_{\xi\eta} + g_{11}\mathbf{x}_{\eta\eta} = 0, \quad \text{s.t.} \ \mathbf{x}|_{\partial\hat{\Omega}} = \partial\Omega, \tag{2}$$

with

$$g_{11}(\mathbf{x}) = \mathbf{x}_\xi \cdot \mathbf{x}_\xi,$$
$$g_{12}(\mathbf{x}) = \mathbf{x}_\xi \cdot \mathbf{x}_\eta,$$
$$g_{22}(\mathbf{x}) = \mathbf{x}_\eta \cdot \mathbf{x}_\eta \tag{3}$$

being the entries of the metric tensor of the mapping (which are nonlinear functions of $\mathbf{x}$). Under certain assumptions of the boundary contour regularity and assuming that $\hat{\Omega}$ is convex, it can be shown that the exact solution of (2) is bijective, justifying a numerical approximation for the purpose of generating a geometry description [1].

EGG has been an established approach in classical meshing for decades and first attempts to apply it to spline-based geometry descriptions were made in [13], where the equations are approximately solved by a collocation at the abscissae of a Gaussian quadrature scheme with cubic Hermite-splines. In [12], the collocation takes place at the Greville-abscissae and the resulting nonlinear equations are solved using a Picard-based iterative scheme, allowing for a wider range of spline-bases. However, as a downside, the consistency order of Greville-based collocation is not optimal. In [9], the equations are discretized with a Galerkin approach and a Newton-based iterative approach is employed for the resulting root-finding problem, allowing for $C^{\geq 1}$-continuous bases. Numerical convergence is accelerated by generating good initial guesses utilizing multigrid-techniques and convergence is typically achieved within 4 (unconstrained) nonlinear iterations.

Unfortunately, none of the aforementioned approaches allow for spline-bases with locally reduced smoothness, limiting their usefullness in practice, since in certain applications C^0-continuities are desirable or unavoidable, notably in multipatch parameterizations or when $\partial\Omega$ is build from a spline-basis with (one or more) p-fold internal knot repetitions (where p refers to the polynomial order of the spline-basis used). To allow for C^0-continuities, one may instead minimize the *Winslow-functional* [16] (whose global minimizer is equal to the exact solution of (2)). Unfortunately, this leads to a formulation in which the Jacobian determinant appears in the denominator, which is why an iterative solution scheme has to be initialized with a bijective initial guess in order to avoid division by zero, restricting it to use cases in which a bijective initial guess is available.

Motivated by our striving for a computationally inexpensive parameterization algorithm that does not have to be initialized by a bijective initial guess and allows for spline-bases with arbitrary continuity properties, in this paper, we augment the discretization proposed in [9] with auxilliary variables, leading to a mixed-FEM type problem. To allow for its application to large-scale problems, we present a solution strategy that tackles the resulting nonlinear root-finding problem with a

Newton-Krylov-based [11] Jacobian-free iterative approach that only operates on the nonlinear part (corresponding to the primary, not auxilliary variables) of the equation. Besides single-patch problems, we will address potential use cases of the algorithm in multipatch settings (in particular with extraordinary vertices), made possible by the support of C^0-continuous spline bases. We conclude this paper with a number of example-parameterizations and a discussion of the results.

2 Problem Formulation

In [9], the following discretization of the governing equations (see Eq. (2)) is proposed:

$$\text{find } \mathbf{x} \in [\text{span } \Sigma]^2 \text{ s.t.}$$

$$\begin{cases} \forall \boldsymbol{\sigma}_i \in [\Sigma_0]^2 : & \int_{\hat{\Omega}} \boldsymbol{\sigma}_i \cdot \mathcal{L}(\mathbf{x}) d\boldsymbol{\xi} = 0 \\ \mathbf{x}|_{\partial \hat{\Omega}} = \partial \Omega \end{cases} \tag{4}$$

where $\Sigma_0 \equiv \{w_i \in \Sigma \mid w_i|_{\partial \hat{\Omega}} = 0\}$.

Similarly, [10] introduces a scaled version of (4), namely:

$$\text{find } \mathbf{x} \in [\text{span } \Sigma]^2 \text{ s.t.}$$

$$\begin{cases} \forall \boldsymbol{\sigma}_i \in [\Sigma_0]^2 : & \int_{\hat{\Omega}} \boldsymbol{\sigma}_i \cdot \tilde{\mathcal{L}}(\mathbf{x}) d\boldsymbol{\xi} = 0 \\ \mathbf{x}|_{\partial \hat{\Omega}} = \partial \Omega \end{cases} \tag{5}$$

where

$$\tilde{\mathcal{L}}(\mathbf{x}) = \frac{\mathcal{L}(\mathbf{x})}{\underbrace{g_{11} + g_{22}}_{\geq 0} + \underbrace{\mu}_{>0}}. \tag{6}$$

Here, $\mu > 0$ is a small positive parameter that is usually taken to be $\mu = 10^{-4}$.

The motivation to solve (5) rather than (4) is based on the observation that numerical root-finding algorithms typically converge faster in this case and that a suitable convergence criterion is less geometry-dependent. Note that the scaling is allowed because the exact solution is unchanged. Therefore, we base our reformulation of the problem on (5).

In order to reduce the highest-order derivatives from two to one, we introduce a new operator in which we replace second order derivatives in $\mathbf{x}$ by the first order derivatives of $\mathbf{u}$ and $\mathbf{v}$, respectively:

$$\mathcal{U}(\mathbf{u}, \mathbf{v}, \mathbf{x}) = \frac{g_{22}\mathbf{u}_\xi - g_{12}\mathbf{u}_\eta - g_{12}\mathbf{v}_\xi + g_{11}\mathbf{v}_\eta}{g_{11} + g_{22} + \mu}. \tag{7}$$

Where $\mathcal{U}$ satisfies

$$\tilde{\mathcal{L}}(\mathbf{x}) = \mathcal{U}(\mathbf{x}_\xi, \mathbf{x}_\eta, \mathbf{x}). \tag{8}$$

A possible reformulation of (5) with auxilliary variables now reads:

find $(\mathbf{u}, \mathbf{v}, \mathbf{x})^T \in [\operatorname{span} \bar{\Sigma}]^4 \times [\operatorname{span} \Sigma]^2$ s.t.

$$\begin{cases} \forall \boldsymbol{\sigma}_i \in [\bar{\Sigma}]^4 \times [\Sigma_0]^2 : \int_{\hat{\Omega}} \boldsymbol{\sigma}_i \cdot \begin{pmatrix} \mathbf{u} - \mathbf{x}_\xi \\ \mathbf{v} - \mathbf{x}_\eta \\ \mathcal{U}(\mathbf{u}, \mathbf{v}, \mathbf{x}) \end{pmatrix} d\boldsymbol{\xi} = 0 \\ \mathbf{x}|_{\partial \hat{\Omega}} = \partial \Omega \end{cases} \tag{9}$$

where $\bar{\Sigma} = \{\bar{w}_1, \ldots, \bar{w}_{\bar{N}}\}$ denotes the basis that is used for the auxilliary variables.

Note that the choice of (7) is not unique. Here, we have chosen to divide $\mathbf{x}_{\xi\eta}$ equally among $\mathbf{u}_\eta$ and $\mathbf{v}_\xi$. In general, any combination

$$\mathbf{x}_{\xi\eta} \to \chi \mathbf{u}_\eta + (1 - \chi) \mathbf{v}_\xi, \tag{10}$$

is valid. Note that since the g_{ij} are functions of $\mathbf{x}_\xi$ and $\mathbf{x}_\eta$, further possible variants are acquired by substituting $\mathbf{u}, \mathbf{v}$ in the g_{ij}.

System (9) now constitutes a discretization of (2) that allows for only C^0-continuous bases at the expense of increasing the problem size from $2|\mathcal{I}_{\text{inner}}|$ to $2|\mathcal{I}_{\text{inner}}| + 4|\bar{\Sigma}|$, where, as before, $\mathcal{I}_{\text{inner}}$ refers to the index set of inner control points.

Let us remark that in certain settings, it suffices to invoke auxilliary variables in one coordinate-direction only. A possible problem formulation for the ξ-direction reads:

find $(\mathbf{u}, \mathbf{x})^T \in [\operatorname{span} \bar{\Sigma}]^2 \times [\operatorname{span} \Sigma]^2$ s.t.

$$\begin{cases} \forall \boldsymbol{\sigma}_i \in [\bar{\Sigma}]^2 \times [\Sigma_0]^2 : \int_{\hat{\Omega}} \boldsymbol{\sigma}_i \cdot \begin{pmatrix} \mathbf{u} - \mathbf{x}_\xi \\ \mathcal{U}^\xi(\mathbf{u}, \mathbf{x}) \end{pmatrix} d\boldsymbol{\xi} = 0 \\ \mathbf{x}|_{\partial \hat{\Omega}} = \partial \Omega \end{cases} \tag{11}$$

with (for instance)

$$\mathcal{U}^\xi(\mathbf{u}, \mathbf{x}) = \frac{g_{22} \mathbf{u}_\xi - g_{12} \mathbf{u}_\eta - g_{12} \mathbf{x}_{\xi\eta} + g_{11} \mathbf{x}_{\eta\eta}}{g_{11} + g_{22} + \mu}. \tag{12}$$

And similarly for the η-direction.

The above approach is useful if C^0-continuities are only required in a single coordinate-direction so that the total number of degrees of freedom (DOFs) can be reduced.

3 Solution Strategy

Systems (9) and (11) are nonlinear and have to be solved with an iterative algorithm. We will discuss a solution algorithm that is loosely based on the Newton-approach proposed in [9]. However, we tweak it in order to reduce computational costs and memory requirements by exploiting many characteristics of the problem at hand. First, we discuss the case in which $\hat{\Omega}$ is given by a single patch, after which we generalize our solution strategy to multipatch-settings (in particular with topologies that contain extraordinary vertices).

3.1 Single Patch Parameterizations

With $\mathbf{x} = \mathbf{x}[\mathbf{c}]$, where $\mathbf{c}$ is a vector containing the $\mathbf{c}_j$ in (1) (while freezing the $\mathbf{c}_i$ that follow from the boundary condition) and $(\mathbf{u}, \mathbf{v})^T = (\mathbf{u}, \mathbf{v})^T[\mathbf{d}]$, where $\mathbf{d} = (\mathbf{d}^u, \mathbf{d}^v)^T$ is a vector containing $\mathbf{d}^u_i$ and $\mathbf{d}^v_i$ in

$$\mathbf{u}[\mathbf{d}^u] = \sum_i \mathbf{d}^u_i \bar{w}_i,$$

$$\mathbf{v}[\mathbf{d}^v] = \sum_i \mathbf{d}^v_i \bar{w}_i, \tag{13}$$

we can reinterpret (9) as a problem in $\mathbf{c}$ and $\mathbf{d}$. It has a residual vector of the form

$$\mathbf{R}(\mathbf{d}, \mathbf{c}) = \begin{pmatrix} R_L(\mathbf{d}, \mathbf{c}) \\ R_N(\mathbf{d}, \mathbf{c}) \end{pmatrix}, \tag{14}$$

where R_L refers to the linear part in (9) (the projection of the auxilliary variables onto $\mathbf{x}_\xi$ and $\mathbf{x}_\eta$) and R_N to the nonlinear (the part involving the operator $\mathcal{U}(\mathbf{u}, \mathbf{v}, \mathbf{x})$).

The Newton-approach from [9] requires the assembly of the Jacobian

$$J_R = \begin{pmatrix} \frac{\partial R_L}{\partial \mathbf{d}} & \frac{\partial R_L}{\partial \mathbf{c}} \\ \frac{\partial R_N}{\partial \mathbf{d}} & \frac{\partial R_N}{\partial \mathbf{c}} \end{pmatrix} \equiv \begin{pmatrix} A & B \\ C & D \end{pmatrix} \tag{15}$$

of (9) at every Newton-iteration. The matrices A and B, corresponding to the linear part in (9), are not a function of $\mathbf{c}$ and $\mathbf{d}$ and thus have to be assembled only once. In fact, A is block-diagonal with blocks given by the parametric mass matrix $\bar{M}$ over the auxilliary basis $\bar{\Sigma} = \{\bar{w}_1, \ldots, \bar{w}_{\bar{N}}\}$ with entries

$$\bar{M}_{ij} = \int_{\hat{\Omega}} \bar{w}_i \bar{w}_j d\xi, \tag{16}$$

while B is block-diagonal with blocks whose columns are given by a subset of the columns of the matrices $\bar{M}^\xi$ and $\bar{M}^\eta$ with entries

$$\bar{M}^\xi_{ij} = \int_{\hat{\Omega}} \bar{w}_i w_{j\xi} d\boldsymbol{\xi} \tag{17}$$

and

$$\bar{M}^\eta_{ij} = \int_{\hat{\Omega}} \bar{w}_i w_{j\eta} d\boldsymbol{\xi}. \tag{18}$$

For given $\mathbf{c}$ and $\mathbf{d}$, the Newton search-direction is computed from a system of the form

$$\begin{pmatrix} A & B \\ C & D \end{pmatrix} \begin{pmatrix} \Delta \mathbf{d} \\ \Delta \mathbf{c} \end{pmatrix} = \begin{pmatrix} \mathbf{a} \\ \mathbf{b} \end{pmatrix}, \tag{19}$$

where $C = C(\mathbf{d}, \mathbf{c})$ and $D = D(\mathbf{d}, \mathbf{c})$ are, unlike A and B, not constant and have to be reassembled in each iteration. We form the Schur-complement of A, in order to yield an equation for $\Delta \mathbf{c}$ only, namely:

$$\underbrace{(D - CA^{-1}B)}_{\tilde{D}} \Delta \mathbf{c} = \mathbf{b} - CA^{-1}\mathbf{a}. \tag{20}$$

In order to avoid the computationally expensive assembly of C and D, we solve (20) with a Newton-Krylov [11] algorithm which only requires the evaluation of vector products $\tilde{D}\mathbf{s}$, which can be approximated with finite differences rather than explicit assembly of C and D. Since

$$C\mathbf{s}_1 + D\mathbf{s}_2 = \frac{R_N(\mathbf{d} + \epsilon \mathbf{s}_1, \mathbf{c} + \epsilon \mathbf{s}_2) - R_N(\mathbf{d}, \mathbf{c})}{\epsilon} + O(\epsilon), \tag{21}$$

we have

$$\tilde{D}\mathbf{s} \simeq \frac{R_N(\mathbf{d} - \epsilon A^{-1}B\mathbf{s}, \mathbf{c} + \epsilon \mathbf{s}) - R_N(\mathbf{d}, \mathbf{c})}{\epsilon}, \tag{22}$$

and

$$CA^{-1}\mathbf{a} \simeq \frac{R_N(\mathbf{d} + \epsilon A^{-1}\mathbf{a}, \mathbf{c}) - R_N(\mathbf{d}, \mathbf{c})}{\epsilon}, \tag{23}$$

for ϵ small. The optimal choice of ϵ is discussed in [11].

We compute products of the form $\mathbf{q} = A^{-1}\mathbf{t}$ from the solution of the system $A\mathbf{q} = \mathbf{t}$, which has for $\mathbf{t} = B\mathbf{s}$ (see Eq. (22)) and $\mathbf{t} = \mathbf{a}$ (see Eq. (23)) the form of a (separable) L_2-projection. Let

$$\mathbf{x}^0[\mathbf{c}] = \sum_{j \in I_{\text{inner}}} \mathbf{c}_j w_j. \tag{24}$$

Product $\mathbf{q} = A^{-1}B\mathbf{s}$ satisfies

$$\mathbf{q} = (\mathbf{q}^u, \mathbf{q}^v)^T = \underset{(\tilde{\mathbf{q}}^u, \tilde{\mathbf{q}}^v)}{\operatorname{argmin}} \frac{1}{2} \int_{\hat{\Omega}} \left\| \begin{bmatrix} \mathbf{u}[\tilde{\mathbf{q}}^u] \\ \mathbf{v}[\tilde{\mathbf{q}}^v] \end{bmatrix} - \begin{bmatrix} \mathbf{x}^0_\xi[\mathbf{s}] \\ \mathbf{x}^0_\eta[\mathbf{s}] \end{bmatrix} \right\|^2 d\boldsymbol{\xi}, \tag{25}$$

and similarly for $\mathbf{q} = A^{-1}\mathbf{a}$.

As such, A is block-diagonal and composed of separable mass matrices $\bar{M} = \bar{m}_\xi \otimes \bar{m}_\eta$

$$A = \begin{pmatrix} \bar{m}_\xi \otimes \bar{m}_\eta & & \\ & \ddots & \\ & & \bar{m}_\xi \otimes \bar{m}_\eta \end{pmatrix}, \tag{26}$$

where $\bar{m}_\xi$ and $\bar{m}_\eta$ refer to the univariate mass matrices resulting from the tensor-product structure of Σ. Therefore, we have

$$A^{-1} = \begin{pmatrix} (\bar{m}_\xi^{-1}) \otimes (\bar{m}_\eta^{-1}) & & \\ & \ddots & \\ & & (\bar{m}_\xi^{-1}) \otimes (\bar{m}_\eta^{-1}) \end{pmatrix}. \tag{27}$$

We follow the methodology from [6], where a computationally inexpensive inversion of this $2D$ mass matrix is achieved by repeated inversion with the $1D$ mass matrices $\bar{m}_\xi$ and $\bar{m}_\eta$. Here, we do direct inversion of the $1D$ mass matrices by computing their Cholesky-decompositions [15]. An inversion can be done in only $O(\bar{N})$ arithmetic operations and Cholesky-decompositions have to be formed only once, thanks to the fact that A is constant.

After solving (20), $\Delta \mathbf{d}$ is found by solving

$$A\Delta\mathbf{d} = \mathbf{a} - B\Delta\mathbf{c}. \tag{28}$$

Upon completion, the vector $\mathbf{n} \equiv (\Delta\mathbf{d}, \Delta\mathbf{c})^T$ constitutes the Newton search-direction. We update the current iterate $(\mathbf{d}, \mathbf{c})^T$ by adding $\nu\mathbf{n}$, where the optimal value of $\nu \in (0, 1]$ is estimated through a line-search routine. Above steps are repeated until the norm of $\mathbf{n}$ is negligibly small. Upon completion, we extract the c-component from the resulting solution vector which contains the inner control points of the mapping operator **x**, while the **d**-component serves no further purpose and can be discarded.

It should be noted that a single matrix-vector product $\tilde{D}\mathbf{s}$ is slightly more expensive than, for instance, $D\mathbf{s}$, due to the requirement to invert A. However, thanks to the separable nature of A, the costs in (22) are dominated by function evaluations in R_L, which implies that a performance quite similar to that of an approach without auxilliary variables can be achieved.

There exist many possible choices of constructing an initial guess for the **c**-component of the iterative scheme. Common choices are algebraic methods, most notably transfinite interpolation [7]. Once the **c**-component has been computed with one of the available methods, a reasonable way to compute the corresponding **d**-part is through a (separable) projection of $\mathbf{x}_\xi$ and $\mathbf{x}_\eta$ onto $\tilde{\Sigma}$.

Slightly superior initial guesses can be generated using multigrid techniques as demonstrated in [9]. The problem is first solved using a coarser basis and an algebraic initial guess, after which the coarse solution vector is prolonged and subsequently used as an initial guess. This is compatible with the techniques discussed in this section. However, instead of prolonging the full coarse solution vector, we only prolong the **c**-component and compute the corresponding **d**-component using an $L_2(\hat{\Omega})$-projection.

3.2 Multipatch

The reformulation with auxilliary variables has a particularly interesting application in multipatch-settings, especially when extraordinary patch vertices are present. Most of the techniques from Sect. 3.1 are readily applicable but there exist subtle differences that shall be outlined in the following.

Let $\hat{\Omega}$ be a multipatch domain, i.e.,

$$\hat{\Omega} = \bigcup_{i=1}^{n} \hat{\Omega}_i. \tag{29}$$

For convenience, let us assume that each $\hat{\Omega}_i$ is an affine transformation of the reference unit square $\tilde{\Omega} = [0,1]^2$ with corresponding mapping $\mathbf{m}_i : \tilde{\Omega} \to \hat{\Omega}_i$, where

$$\mathbf{m}_i(\mathbf{s}) = A_i \mathbf{s} + \mathbf{b}_i. \tag{30}$$

Here, A_i is an invertible matrix, $\mathbf{b}_i \in \mathbb{R}^2$ some translation and the vector $\mathbf{s} = (s, t)^T$ contains the free variables in $\tilde{\Omega}$. The automated generation of a multipatch structure is a nontrivial task, which is not discussed in this paper. For an overview of possible segmentation techniques, we refer to [3, 17, 5].

Let $\tilde{\mathbf{x}} : \hat{\Omega} \to \Omega$ be such that $\tilde{\mathbf{x}}^{-1} : \Omega \to \hat{\Omega}$ is a harmonic mapping. Assuming that the $\hat{\Omega}_i$ are arranged such that $\hat{\Omega}$ is convex, Rado's theorem [14] applies and a harmonic $\tilde{\mathbf{x}}^{-1}$ is bijective.

In the case of a multipatch domain, pairs of faces $(\gamma_i^\alpha, \gamma_j^\beta) \subset \partial\hat{\Omega}_i \times \partial\hat{\Omega}_j$ and sets of vertices $\{\mathbf{p}_i, \ldots, \mathbf{p}_l\} \subset \partial\hat{\Omega}_i \times \ldots \times \partial\hat{\Omega}_l$ may coincide on $\hat{\Omega}$. As such, the bases Σ and $\bar{\Sigma}$, whose elements constitute single-valued functions on $\hat{\Omega}$ are constructed from the patchwise discontinuous local bases Σ_i and $\bar{\Sigma}_i$ with appropriate degree of freedom (DOF) coupling that canonically follows from the connectivity properties of the $\hat{\Omega}_i$. In the multipatch case, we solve (9) by evaluating the associated integrals through a set of pull backs of the $\hat{\Omega}_i \subset \hat{\Omega}$ into the reference domain $\tilde{\Omega}$. Thanks to the affine nature of the pull back, replacement of $\boldsymbol{\xi}$-derivatives by local $\mathbf{s}$-derivatives is straightforward.

As such, the solution of (9) yields a collection of mappings $\{\mathbf{x}_i\}_i$, with $\mathbf{x}_i : \tilde{\Omega} \to \Omega_i \subset \Omega$, where each $\mathbf{x}_i$ satisfies

$$\mathbf{x}_i \simeq \tilde{\mathbf{x}}|_{\hat{\Omega}_i} \circ \mathbf{m}_i. \tag{31}$$

As the right hand side of (31) is a composition of bijective mappings, the bijectivity of $\mathbf{x}_i$ depends on the quality of the approximation. If the $\mathbf{x}_i$ are bijective, they jointly form a parameterization of Ω.

Unlike in the single-patch setting, the $L_2(\hat{\Omega})$-projection associated with the linear part of the residual vector is not separable. As such, the evaluation of vector products $A^{-1}B\mathbf{s}$ (see Eq. (22)) becomes more involved. A possible workaround is explicit assembly and inversion of the Jacobian of the system (see Eq. (19)), leading to increased computational times and memory requirements.

A possible alternative is the approximation of products of the form $A^{-1}B\mathbf{s}$ by a sequence of patchwise separable operations. In the following, we sketch a plausible approach.

Similar to the single-patch case, products of the form $(\mathbf{q}^u, \mathbf{q}^v)^T = A^{-1}B\mathbf{s}$ satisfy

$$(\mathbf{q}^u, \mathbf{q}^v)^T = \underset{(\tilde{\mathbf{q}}^u, \tilde{\mathbf{q}}^v)^T}{\operatorname{argmin}} \sum_{i=1}^n \frac{1}{2} \int_{\hat{\Omega}_i} \left\| \begin{bmatrix} u[\tilde{\mathbf{q}}^u] \\ v[\tilde{\mathbf{q}}^v] \end{bmatrix} - \begin{bmatrix} x_\xi^0[\mathbf{s}] \\ x_\eta^0[\mathbf{s}] \end{bmatrix} \right\|^2 d\boldsymbol{\xi}. \tag{32}$$

Let

$$\tilde{\Sigma} = \bigcup_{i=1}^n \bar{\Sigma}_i \equiv \{\tilde{w}_i\}_i \tag{33}$$

be the patchwise discontinuous union of local (auxilliary variable) bases and let

$$\tilde{\mathbf{u}}[\mathbf{g}] = \sum_i \mathbf{g}_i \tilde{w}_i,$$

$$\tilde{\mathbf{v}}[\mathbf{h}] = \sum_i \mathbf{h}_i \tilde{w}_i. \tag{34}$$

In order to approximate $(\mathbf{q}^u, \mathbf{q}^v)^T$, we first find

$$(\mathbf{g}, \mathbf{h})^T = \underset{(\tilde{\mathbf{g}}, \tilde{\mathbf{h}})^T}{\operatorname{argmin}} \sum_{i=1}^n \frac{1}{2} \int_{\hat{\Omega}_i} \left\| \begin{bmatrix} \tilde{\mathbf{u}}[\tilde{\mathbf{g}}] \\ \tilde{\mathbf{v}}[\tilde{\mathbf{h}}] \end{bmatrix} - \begin{bmatrix} \mathbf{x}_\xi^0[\mathbf{s}] \\ \mathbf{x}_\eta^0[\mathbf{s}] \end{bmatrix} \right\|^2 d\xi. \tag{35}$$

We perform a patchwise pullback of the L_2-projections into the reference domain where they are solved with the techniques from Sect. 3.1. Thanks to the affine nature of the pullback, the geometric factor associated with $\hat{\Omega}_i$ is constant and given by

$$\det J_i = \det A_i. \tag{36}$$

Therefore, separability is not lost and the same efficiency as in the single-patch case is achieved. We restrict the solution of (35) to $\bar{\Sigma}$ by performing a weighted sum of components that coincide under coupling. Let $\bar{w}_i \in \bar{\Sigma}$ result from a coupling of $\{\tilde{w}_\alpha, \ldots, \tilde{w}_\gamma\} \subset \tilde{\Sigma}$ and let $\{\det J_\alpha, \ldots, \det J_\gamma\}$ denote the set of corresponding local geometric factors. If the $\{\tilde{w}_\alpha, \ldots, \tilde{w}_\gamma\}$ receive control points $\mathbf{g}_\alpha, \ldots, \mathbf{g}_\gamma$ under the projection, we set

$$\mathbf{q}_i^u = \frac{\det J_\alpha \mathbf{g}_\alpha + \ldots + \det J_\gamma \mathbf{g}_\gamma}{\det J_\alpha + \ldots + \det J_\gamma}, \tag{37}$$

and similarly for $\mathbf{q}^v$. Relation (37) induces a canonical restriction operator from span $\tilde{\Sigma}$ to span $\bar{\Sigma}$ that is used to compute $(\mathbf{q}^u, \mathbf{q}^v)^T$ from $(\mathbf{g}, \mathbf{h})^T$.

4 Numerical Experiments

In the following, we present several numerical experiments, demonstrating the functioning of the proposed algorithm. First, we present two single-patch problems after which we present a more involved multipatch parameterization.

In all cases, the auxiliary basis $\bar{\Sigma}$ results from one global h-refinement of the primal basis Σ.

4.1 L-Bend

As a proof of concept, we present results for the well-known single-patch L-bend problem. Wherever possible, we shall compare the results to a direct minimization of the Winslow-functional

$$W(\mathbf{x}) = \int_{\hat{\Omega}} \frac{g_{11} + g_{22}}{\det J} d\boldsymbol{\xi}, \qquad (38)$$

whose global minimizer (over [span Σ]^2) coincides with a numerical approximation of the solution of (2) in the limit where $N \to \infty$ [1]. For the L-bend problem, we employ uniform cubic ($p = 3$) knot-vectors in both directions with a p-fold knot-repetition at $\xi = 0.5$ in order to properly resolve the C^0-continuity. As such we solve (11) rather than (9). Figure 1 shows the resulting parameterization along with the element boundaries under the mapping. The Schur-complement solver converges after 3 iterations which amounts to 106 evaluations of R_N. As can be seen in the figure, the parameterization is symmetric across the line connecting the upper and lower C^0-continuities which is expected behaviour from the shape of the geometry. We regard this as a positive sanity check for the functioning of

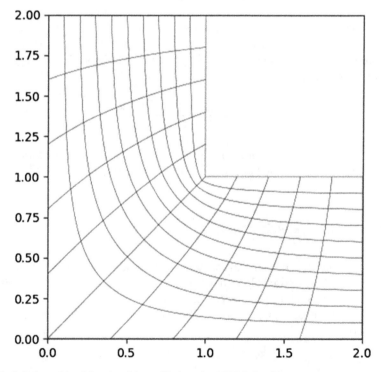

Fig. 1 Solution of the L-bend problem with the mixed-FEM algorithm

the algorithm. Another observation is that despite the presence of knot-repetitions at $\xi = 0.5$, the parameterization shows a large degree of smoothness along the corresponding isoline. Again, this is a positive result since the solution is expected to be an approximation of the global minimizer of (38) (over $\mathbf{x} \in [\text{span } \Sigma]^2$), which, in turn, approximates a smooth function. A substitution of the solution vector $\mathbf{c}_{\text{mf}}$ of the system of Eqs. (11) in (38) gives

$$W(\mathbf{c}_{\text{mf}}) \simeq 3.01518, \tag{39}$$

whereas the global minimizer $\mathbf{c}_W$ of (38) over the same basis yields

$$W(\mathbf{c}_W) \simeq 3.01425. \tag{40}$$

This constitutes another positive sanity check as the results are very close, while a substitution of the PDE-solution is slightly above the global minimum. As such, the PDE-solution comes with all the undesired characteristics of EGG-schemes such as the tendency to yield bundled/spread isolines near concave/convex corners. This does not occur in parameterizations based on the techniques of [8] (see Fig. 2).

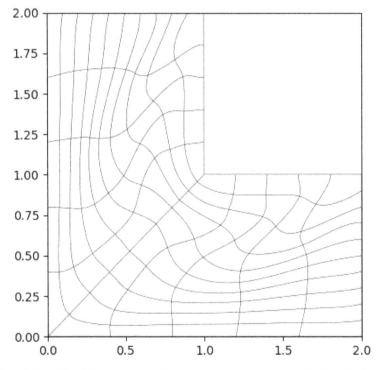

Fig. 2 Solution of the L-bend problem with constrained minimization of the *Area Orthogonality* functional (see [8])

However, the L-bend example is rather contrived since a good parameterization is easily constructed with algebraic techniques. Here, the results only serve as a proof of concept.

4.2 Tube-Like Shaped Geometry

In many cases, segmentation along knots with p-fold repetition and continuation with, for instance, techniques from [9] on the smaller pieces is a viable choice. However, in some cases, a segmentation curve along which to split the geometry into smaller parts may be hard to find. One such example is depicted in Fig. 3 (left), which is a geometry taken from the practical application of numerically simulating a twin-screw machine. For convenience, the $\xi = 0.5$ isoline, across which the mapping is C^0-continuous, has been plotted in red. The usefullness of the proposed algorithm becomes apparent in this case: instead of having to generate a valid $\xi = 0.5$ isoline, the isoline establishes itself from the solution of the PDE-problem.

As in the L-bend problem, we observe that the resulting parameterization exhibits a great degree of smoothness across the $\xi = 0.5$ isoline, despite the continuity properties of Σ and the spiked upper and lower boundaries.

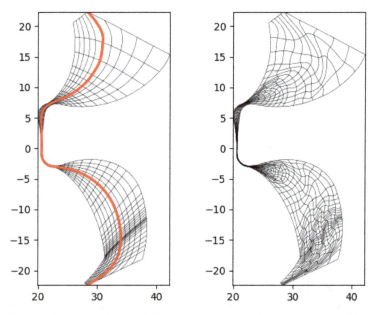

Fig. 3 PDE-based parameterization (left) and area-orthogonality minimized parameterization (right) of a tube-like shaped geometry

The proposed algorithm produces superior results to the constrained optimization approach from [8] (see Fig. 3, right). In fact, here we initialized the optimization by the PDE-solution, as the solver struggles to find a feasible initial guess through optimization. This confirms the finding from [9] that EGG-based approaches may be a viable alternative to finding feasible initial guesses for approaches based on optimization. Furthermore, we note the striking difference in the required number of iterations, which amount to over 100 (constrained) in the optimization, while the PDE-solver converges in only 7 iterations.

The poor performance of the optimization-approach can be explained by tiny gaps contained in the geometry, leading to natural jumps in the magnitude of the Jacobian determinant. As most cost functions are functions of the g_{ij}, they are very sensitive to jumps in det J. This is further evidenced by the poor grid quality in the narrow part of the geometry (see Fig. 4 right). In our experience, this is not the case for the PDE-solution (see Fig. 4 left) and we successfully employed the approach for the automatic generation of a large number of similar geometries.

Finally, it should be noted that a comparison to the global minimizer of the Winslow-energy is not possible since the gradient-based optimizer we employed failed to further reduce the cost function from the evaluation of the PDE-solution.

4.3 Multipatch Problem: The Bat Geometry

Another interesting application of the proposed algorithm is that of a multipatch parameterization. In Sect. 4.2, we have successfully employed the algorithm to a geometry with a C^0-continuity along the $\xi = 0.5$ isoline, which might as well be regarded as a two-patch parameterization with coupling along aforementioned isoline. A much more interesting multipatch application would be that of an uneven number of patches with extraordinary vertices. We are considering the diamond-shaped triple-patch domain depicted in Fig. 5, left. The target boundaries form the

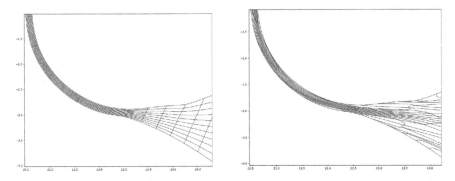

Fig. 4 Zoom-in on the PDE-based parameterization and area-orthogonality minimization parameterization

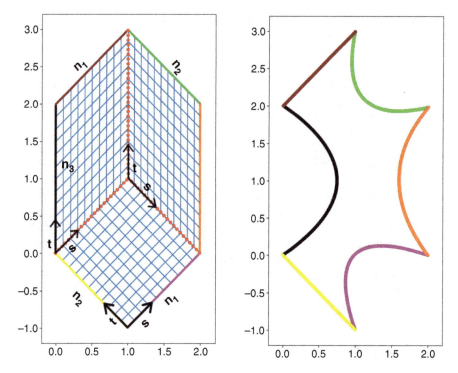

Fig. 5 Diamond shaped multipatch domain (left) and the target boundaries (right). Here, $n_1 = 10$, $n_2 = 11$ and $n_3 = 12$ denote the number of (uniformly-spaced) elements in each coordinate direction. There are no internal knot repetitions

bat-shaped contour depicted in Fig. 5, right. Note that, as required in Sect. 3.2, the domain forms a convex subset of $\mathbb{R}^2$. For convenience we have highlighted the positions of the various boundaries under the mapping in different colors. Of course, of major interest shall be how the dotted red curve(s) in Fig. 5 (left) are deformed under the mapping. Figure 6 (left) shows the mapping we utilize to initialize the Newton-Krylov solver while Fig. 6 (right) shows the resulting geometry. Even though better initial guesses are easily constructed, here we have chosen to initialize the solver with a folded initial guess in order to demonstrate that bijectivity is not a necessary condition for convergence. The Newton-Krylov solver converges after 6 nonlinear iterations. The dotted red curves in Fig. 6 (right) show the internal interfaces of $\hat{\Omega}$ under the mapping. We see that the patch interfaces are mapped into the interior of Ω. The resulting geometry is bijective. However, the isolines make steep angles by the internal patch interfaces. This results from the additional pull back of $\tilde{\mathbf{x}}|_{\hat{\Omega}_i}$ into $\tilde{\Omega}$ via the operator $\mathbf{m}_i$ (see Eq. (31)), which generally introduces a C^0-continuity in the composite mapping. Higher-order smoothness across patch interfaces is generally difficult to achieve and usually done by constructing bases whose elements possess higher-order continuity sufficiently far away from the

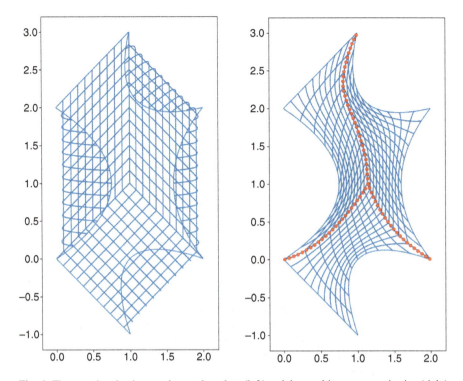

Fig. 6 The mapping that is passed on to the solver (left) and the resulting parameterization (right)

extraordinary vertices. However, note that such bases may not allow for patchwise-affine transformations such that $L_2(\hat{\Omega}_i)$-projections lose their separability property. For a more rigorous definition of smoothness on multipatch topologies and strategies to build bases with local $C^{\geq 1}$ smoothness on patch interfaces, we refer to [4].

5 Conclusion

We have formulated an IgA-suitable EGG-algorithm that is compatible with spline bases Σ possessing arbitrary continuity properties (whereby *arbitrary* stands for global $C^{\geq 0}$-continuity) by introducing a set of auxiliary variables. We proposed an iterative Newton-Krylov approach operating on the Schur-complement of the linear part of the resulting nonlinear system of equations, which operates efficiently and reduces memory requirements. As such, it is suitable for large problems. Unlike similar C^0-compatible EGG-based approaches, the iterative solution method does not have to be initialized with a bijective mapping, significantly improving its usability in practice. However, this major advantage comes at the expense of increasing the problem size from N to $\simeq cN$, where $c = 2$ or $c = 3$, depending on the context. The impact is partially mitigated by the specially-taylored iterative solution algorithm.

We have presented three numerical experiments, two with a single patch and one resulting from a triple-patch configuration. In the single-patch case, we concluded that a substitution of the PDE-solution into the Winslow functional (Eq. (38)) yields an outcome that is close to that of the global function-minimizer (which is generally hard to find through direct minimization, due to the presence of $\det J$ in the denominator of Eq. (38)). As such, we concluded that the algorithm operates as expected and offers a viable alternative to direct minimization of (38). However, it also comes with all the known drawbacks of EGG-based approaches and the two single-patch test cases demonstrate that it can yield inferior and superior results to other techniques, depending on the characteristics of the geometry.

As convergence is typically reached within only a few iterations, we conclude that the algorithm can serve as a computationally inexpensive method to initialize other methods that require a bijective initial guess. The required number of iterations can be further reduced by employing multigrid techniques (see [9]) but this has not been implemented yet.

A major use case of the proposed algorithm is that of multipatch applications. In Sect. 4.3, we presented results of the application to a triple-patch topology, where we successfully generated a patchwise bijective parameterization by approximating the composition of an inverse-harmonic mapping and patchwise affine transformations. The position of internal patch-interfaces under the mapping do not have to be imposed manually but follow naturally from the composite PDE-solution.

Finally, we observed that the composition with affine transformations results in nonsmooth transitions at patch interfaces. Higher-order smoothness can be achieved by a clever coupling of inter-patch DOFs sufficiently far away from extraordinary vertices.

References

1. B. N. Azarenok. Generation of structured difference grids in two-dimensional nonconvex domains using mappings. *Computational Mathematics and Mathematical Physics*, 49(5):797–809, 2009.
2. L. T. Biegler and V. M. Zavala. Large-scale nonlinear programming using IPOPT: An integrating framework for enterprise-wide dynamic optimization. *Computers & Chemical Engineering*, 33(3):575–582, 2009.
3. F. Buchegger and B. Jüttler. Planar multi-patch domain parameterization via patch adjacency graphs. *Computer-Aided Design*, 82:2–12, 2017.
4. F. Buchegger, B. Jüttler, and A. Mantzaflaris. Adaptively refined multi-patch B-splines with enhanced smoothness. *Applied Mathematics and Computation*, 272:159–172, 2016.
5. A. Falini and B. Jüttler. Thb-splines multi-patch parameterization for multiply-connected planar domains via template segmentation. *Journal of Computational and Applied Mathematics*, 349:390–402, 2019.

6. L. Gao and V. M. Calo. Fast isogeometric solvers for explicit dynamics. *Computer Methods in Applied Mechanics and Engineering*, 274:19–41, 2014.
7. W. J. Gordon and C. A. Hall. Transfinite element methods: Blending-function interpolation over arbitrary curved element domains. *Numerische Mathematik*, 21(2):109–129, 1973.
8. J. Gravesen, A. Evgrafov, D.-M. Nguyen, and P. Nørtoft. Planar parametrization in isogeometric analysis. In *International Conference on Mathematical Methods for Curves and Surfaces*, pages 189–212. Springer, 2012.
9. J. Hinz, M. Möller, and C. Vuik. Elliptic grid generation techniques in the framework of isogeometric analysis applications. *Computer Aided Geometric Design*, 2018.
10. J. Hinz, M. Möller, and C. Vuik. Spline-based parameterization techniques for twin-screw machine geometries. In *IOP Conference Series: Materials Science and Engineering*, volume 425, page 012030. IOP Publishing, 2018.
11. D. A. Knoll and D. E. Keyes. Jacobian-free Newton–Krylov methods: A survey of approaches and applications. *Journal of Computational Physics*, 193(2):357–397, 2004.
12. P. Lamby and K. Brakhage. Elliptic grid generation by B-spline collocation. In *Proceedings of the 10th International Conference on Numerical Grid Generation in Computational Field Simulations, FORTH, Crete, Greece*, 2007.
13. J. Manke. A tensor product B-spline method for numerical grid generation. Technical report, Washington Univ., Seattle, WA (USA). Dept. of Applied Mathematics, 1989.
14. R. M. Schoen and S.-T. Yau. *Lectures on harmonic maps*, volume 2. Amer Mathematical Society, 1997.
15. M. C. Seiler and F. A. Seiler. Numerical recipes in C: The art of scientific computing. *Risk Analysis*, 9(3):415–416, 1989.
16. A. M. Winslow. Adaptive-mesh zoning by the equipotential method. Technical report, Lawrence Livermore National Lab., CA (USA), 1981.
17. S. Xiao, H. Kang, X.-M. Fu, and F. Chen. Computing iga-suitable planar parameterizations by polysquare-enhanced domain partition. *Computer Aided Geometric Design*, 62:29–43, 2018.

Preconditioning for Linear Systems Arising from IgA Discretized Incompressible Navier–Stokes Equations

Hana Horníková and Cornelis Vuik

Abstract We deal with efficient techniques for numerical simulation of the incompressible fluid flow based on the Navier–Stokes equations discretized using the isogeometric analysis approach. Typically, the most time-consuming part of the simulation is solving the large saddle-point type linear systems arising from the discretization. These systems can be efficiently solved by Krylov subspace methods, but the choice of the preconditioner is crucial.

In our study we test several preconditioners developed for the incompressible Navier–Stokes equations discretized by a finite element method, which can be found in the literature. We study their efficiency for the linear systems arising from the IgA discretization, where the matrix is usually less sparse compared to those from finite elements.

Our aim is to develop a fast solver for a specific problem of flow in a water turbine. It brings several complications like periodic boundary conditions at nonparallel boundaries and computation in a rotating frame of reference. This makes the system matrix even less sparse with a more complicated sparsity pattern.

1 Introduction

This work is motivated by numerical simulation of the incompressible fluid flow modeled by the Navier–Stokes/RANS equations with the aim of automatic shape optimization of runner blades of a water turbine. The governing equations are discretized using the isogeometric analysis (IgA) approach. The IgA approach has many common features with the finite element analysis (FEA). The main difference

---

H. Horníková (✉)
Faculty of Applied Sciences, University of West Bohemia, Plzeň, Czech Republic
e-mail: hhornik@kma.zcu.cz

C. Vuik
Delft Institute of Applied Mathematics, Delft University of Technology, Delft, The Netherlands
e-mail: c.vuik@tudelft.nl

is higher smoothness of the solution yielding higher accuracy per degree of freedom than standard finite elements with basis functions of the same order. Another advantage, which is important in the context of incompressible flows, is that it is possible to construct divergence conforming discretization spaces for complex domains using the isogeometric generalization of Raviart-Thomas elements [7, 8]. IgA is also suitable for the purpose of shape optimization, because it allows us to represent the domain boundaries exactly. Similarly to FEA, the IgA discretization of the linearized Navier–Stokes/RANS equations using an LBB stable pair of solution spaces leads to a sequence of sparse saddle-point type linear systems.

The solution of these linear systems represents the main bottleneck of the simulation process. The exact solution is unrealizable for large real world problems, because of very high time and memory requirements of the direct solvers. A promising approach to the iterative solution of these systems is the combination of Krylov subspace methods for nonsymmetric matrices with so-called block triangular preconditioners or SIMPLE-type preconditioners, both based on splitting the system into a velocity and a pressure part. As examples of block triangular preconditioners, we name the pressure convection-diffusion preconditioner (PCD) proposed by Kay, Loghin, Wathen in [12], the least-squares commutator (LSC) preconditioner by Elman et al. [4, 6] or the augmented Lagrangian preconditioner by Benzi and Olshanskii [1]. For an overview of the SIMPLE-type preconditioners, see e.g. [18].

In this paper, we present results of some numerical experiments for GMRES with these preconditioners applied to several linear systems arising from IgA discretization of the incompressible Navier–Stokes equations. We observe some of their properties like dependence of the convergence on the mesh refinement or the Reynolds number for a classical benchmark problem of flow in a 2D backward facing step domain. We also test their performance for a simple 2D problem with periodic boundary conditions on nonparallel sides and a mesh refined locally along these sides, which mimics some of the typical aspects of computations in the water turbine, since the turbine domain is radially symmetric and we usually need to refine the mesh near the blades. Although these 2D domains are very simple, they are intentionally described using higher degree B-splines which is also typical for the turbine geometries.

The structure of this paper is as follows. In Sect. 2 the discretization of the unsteady incompressible Navier–Stokes equations is given and the structure of the matrices obtained from the discretization of the problem in the water turbine is described in more detail. Section 3 gives a brief overview of selected preconditioners. In Sect. 4 we present the results of the numerical experiments and we conclude with Sect. 5.

2 Problem Formulation

The mathematical simulation of incompressible viscous Newtonian flow is based on the incompressible Navier–Stokes equations (NSE). Let $\Omega \subset \mathbb{R}^d$ be a bounded domain, d being the number of spatial dimensions, with the boundary $\partial\Omega$ consisting of two complementary parts, Dirichlet $\partial\Omega_D$ and Neumann $\partial\Omega_N$, and $T > 0$ is an upper bound of the time interval of interest $[0, T]$. The initial boundary value incompressible Navier–Stokes problem is given as a system of $d + 1$ equations together with initial conditions and mixed boundary conditions

$$
\begin{aligned}
\frac{\partial \mathbf{u}}{\partial t} + (\mathbf{u} \cdot \nabla)\mathbf{u} - \nu \Delta \mathbf{u} + \nabla p &= \mathbf{0} & \text{in } \Omega \times [0, T], \\
\nabla \cdot \mathbf{u} &= 0 & \text{in } \Omega \times [0, T], \\
\mathbf{u}(\mathbf{x}, 0) &= \mathbf{u}_0(\mathbf{x}) & \text{in } \Omega, \\
\mathbf{u} &= \mathbf{g}_D & \text{on } \partial\Omega_D, \\
\nu \frac{\partial \mathbf{u}}{\partial \mathbf{n}} - \mathbf{n} p &= \mathbf{0} & \text{on } \partial\Omega_N,
\end{aligned}
\tag{1}
$$

where $\mathbf{u}$ is the flow velocity, p is the kinematic pressure, ν is the kinematic viscosity and $\mathbf{u}_0$, $\mathbf{g}_D$ are given functions. Note that the condition $\nu \frac{\partial \mathbf{u}}{\partial \mathbf{n}} - \mathbf{n} p = \mathbf{0}$ represents the classical "do-nothing" boundary condition resulting from the weak formulation of the momentum equations in (1). It does not have a physical meaning, but it is suitable at artificial outflow boundaries when modeling flows through a truncated domain, assuming that the physical domain continues further (see e.g. [9, 11] where different outflow boundary conditions for such problems are discussed).

2.1 Discretization and Linearization

The isogeometric analysis approach is based on the Galerkin method. One of the approaches to the discretization of time-dependent problems is to discretize the time derivative using finite differences first, arriving to a set of spatial problems that can be discretized using the Galerkin method. Using a backward finite difference with time step Δt, we obtain the following set of equations

$$
\begin{aligned}
\frac{\mathbf{u}^{n+1} - \mathbf{u}^n}{\Delta t} + (\mathbf{u}^{n+1} \cdot \nabla)\mathbf{u}^{n+1} - \nu \Delta \mathbf{u}^{n+1} + \nabla p^{n+1} &= \mathbf{0} & \text{in } \Omega, \\
\nabla \cdot \mathbf{u}^{n+1} &= 0 & \text{in } \Omega.
\end{aligned}
\tag{2}
$$

The weak formulation is as follows: find $\mathbf{u}^{n+1} \in V$ and $p^{n+1} \in L_2(\Omega)$ such that

$$\frac{1}{\Delta t}\int_\Omega \mathbf{u}^{n+1}\cdot\mathbf{v} + \nu\int_\Omega \nabla\mathbf{u}^{n+1}:\nabla\mathbf{v} + \int_\Omega (\mathbf{u}^{n+1}\cdot\nabla\mathbf{u}^{n+1})\cdot\mathbf{v}$$
$$-\int_\Omega p^{n+1}\nabla\cdot\mathbf{v} = \frac{1}{\Delta t}\int_\Omega \mathbf{u}^n\cdot\mathbf{v},$$
$$\int_\Omega q\nabla\cdot\mathbf{u}^{n+1} = 0,$$
(3)

for all $\mathbf{v} \in V_0$ and $q \in L_2(\Omega)$, where

$$V = \{\mathbf{u} \in H^1(\Omega)^d \mid \mathbf{u} = \mathbf{g}_D \text{ on } \partial\Omega_D\},$$
$$V_0 = \{\mathbf{v} \in H^1(\Omega)^d \mid \mathbf{v} = \mathbf{0} \text{ on } \partial\Omega_D\}.$$
(4)

After discretization using a stable pair of discrete solution spaces (e.g. the isogeometric generalization of Taylor–Hood, Nédélec or Raviart-Thomas elements [3]) and linearization of the convective term, we get a sequence of saddle-point type linear systems of the form

$$\begin{bmatrix} F & B^T \\ B & 0 \end{bmatrix}\begin{bmatrix} u \\ p \end{bmatrix} = \begin{bmatrix} f \\ g \end{bmatrix},$$
(5)

where $F \in \mathbb{R}^{d\cdot n_u \times d\cdot n_u}$ is block diagonal (in case of Picard linearization, which is used in this paper) with the diagonal blocks containing the discretization of the convection-diffusion operator and the term coming from the discretized time derivative on the left-hand side of (3). The matrices $B^T \in \mathbb{R}^{d\cdot n_u \times n_p}$ and $B \in \mathbb{R}^{n_p \times d\cdot n_u}$ are discrete gradient and negative divergence operators, respectively, and n_u, n_p denote the number of velocity and pressure unknowns. The right-hand side of (5) contains the remaining part of the discretized time derivative and the eliminated Dirichlet boundary conditions.

2.2 Motivational Problem

As mentioned above, this work is motivated by flow simulation in water turbines. It involves two phenomena, which influence the matrix structure of the resulting linear system: periodic boundary conditions and a rotating frame of reference in the runner wheel.

In Fig. 1 we show an example of a Kaplan turbine geometry (left picture) with the computational domains for the stationary and rotating part (right top and

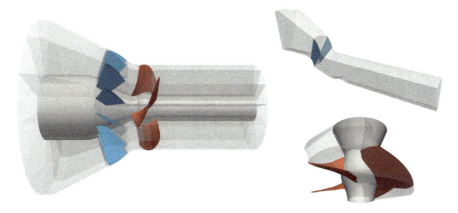

Fig. 1 The Kaplan turbine geometry (left) and particular computational domains for stationary part (right top) and rotating part (right bottom)

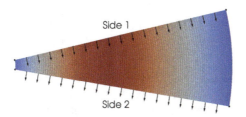

Fig. 2 Velocity field in a cross-section of a periodic 3D domain

bottom picture). Since both stationary and rotating part are radially periodic, we can define each computational domain as a strip between two vanes/blades with periodic conditions.

Since the periodic sides are not parallel and velocity is a vector quantity, we cannot simply identify the degrees of freedom on both sides. In Fig. 2 we display a cross-section of the stationary domain with velocity direction vectors along the periodic sides. It can be seen that instead of just copying the vector from side 1 to side 2, we have to rotate it first.

Without loss of generality, assume that the axis of the turbine is identified with the x-axis. Then the relation between corresponding velocity vectors is

$$\begin{bmatrix} u_1^1 \\ u_2^1 \\ u_3^1 \end{bmatrix} = \begin{bmatrix} 1 & 0 & 0 \\ 0 & \cos\varphi & -\sin\varphi \\ 0 & \sin\varphi & \cos\varphi \end{bmatrix} \begin{bmatrix} u_1^2 \\ u_2^2 \\ u_3^2 \end{bmatrix}, \qquad (6)$$

where u_i^1, $i = 1, 2, 3$, are velocity components on side 1 and u_i^2, $i = 1, 2, 3$, are velocity components on side 2, $\varphi = 2\pi/n_b$ and n_b is the number of blades/vanes.

The application of the periodic conditions can be done by discretization of the given problem with no boundary conditions at the periodic sides, arriving to a linear system with the structure

$$\begin{bmatrix} A & 0 & 0 & B_1^T \\ 0 & A & 0 & B_2^T \\ 0 & 0 & A & B_3^T \\ B_1 & B_2 & B_3 & 0 \end{bmatrix} \begin{bmatrix} u_1 \\ u_2 \\ u_3 \\ p \end{bmatrix} = \begin{bmatrix} f_1 \\ f_2 \\ f_3 \\ g \end{bmatrix}, \qquad (7)$$

and adding multiples of the rows and columns corresponding to the periodic side 2 to the rows and columns corresponding to the periodic side 1 using the transformation (6) and deleting them from the system. This leads to the modified system

$$\begin{bmatrix} \tilde{A} & 0 & 0 & \tilde{B}_1^T \\ 0 & \hat{A} & -C & \tilde{B}_2^T \\ 0 & C & \hat{A} & \tilde{B}_3^T \\ \tilde{B}_1 & \tilde{B}_2 & \tilde{B}_3 & 0 \end{bmatrix} \begin{bmatrix} \tilde{u}_1 \\ \tilde{u}_2 \\ \tilde{u}_3 \\ \tilde{p} \end{bmatrix} = \begin{bmatrix} \tilde{f}_1 \\ \tilde{f}_2 \\ \tilde{f}_3 \\ \tilde{g} \end{bmatrix}, \qquad (8)$$

which is smaller than the original system by the number of degrees of freedom (DOFs) on side 2. After solving the system, the solution coefficients corresponding to side 2 can be obtained by rotating the coefficients corresponding to side 1 back by angle $-\varphi$.

Another nonzero off-diagonal blocks are added to the system matrix in the runner wheel domain, i.e. considering the rotating frame of reference. However, since this paper only presents results of some of the first numerical experiments on simpler domains which do not involve rotation, we do not go into details here.

3 Solution Methods

The linear systems resulting from the discretization can be solved either directly or iteratively. Direct solvers are applicable only for relatively small systems because of their very high time and memory requirements. In practice, we usually deal with very large systems, therefore an efficient iterative method is needed. Among iterative methods, Krylov subspace methods are the most commonly used in applications and can be very efficient if combined with a good preconditioning technique. Since our matrices are nonsymmetric, we have to use a Krylov subspace method for nonsymmetric matrices. The most popular ones are GMRES (generalized minimum residual) and BiCGSTAB (biconjugate gradient stabilized).

A good preconditioner for a Krylov subspace method should be such that the preconditioned matrix has a low degree minimal polynomial, which implies a low maximal dimension of the generated Krylov subspace. In other words, it is desirable that the preconditioned matrix has only a few distinct eigenvalues and is diagonalizable or at least its Jordan canonical form has only small Jordan blocks.

3.1 Block Triangular Preconditioners

A popular class of preconditioners for the saddle-point type problems are the block triangular preconditioners based on splitting the system into a velocity and a pressure part, developed for finite element discretizations of the Navier–Stokes equations. An overview of these preconditioners can be found e.g. in [17]. Their construction is based on the block LDU decomposition of the system matrix in (5)

$$\begin{bmatrix} F & B^T \\ B & 0 \end{bmatrix} = \begin{bmatrix} I & 0 \\ BF^{-1} & I \end{bmatrix} \begin{bmatrix} F & 0 \\ 0 & S \end{bmatrix} \begin{bmatrix} I & F^{-1}B^T \\ 0 & I \end{bmatrix}, \quad (9)$$

where $S = -BF^{-1}B^T$ is the Schur complement matrix. This suggests the following choice of the preconditioner matrix

$$P = \begin{bmatrix} F & B^T \\ 0 & S \end{bmatrix}. \quad (10)$$

Then the right preconditioned matrix has obviously all eigenvalues equal to one, since it is a lower triangular matrix with all ones on its main diagonal. As shown in [15], the minimal polynomial of the preconditioned matrix is of degree 2 and therefore GMRES converges in at most 2 iterations. The same holds for left preconditioning.

The computation of $P^{-1}r$ is performed by solving the linear system

$$\begin{bmatrix} F & B^T \\ 0 & S \end{bmatrix} \begin{bmatrix} z_u \\ z_p \end{bmatrix} = \begin{bmatrix} r_u \\ r_p \end{bmatrix} \quad (11)$$

in the steps summarized in Algorithm 1.

Algorithm 1 Application of P^{-1}

1: Solve $Sz_p = r_p$
2: Update $r_u = r_u - B^T z_p$
3: Solve $Fz_u = r_u$

In practice, the system with F is usually solved approximately, e.g. by a small number of iterations of some iterative method or one or more V-cycles of a multigrid solver for convection-diffusion equations. The solution of the system with the Schur complement is not that straightforward. We do not construct S explicitly, because it would require the explicit construction of F^{-1}, since it is multiplied with rectangular matrices from both sides. Furthermore, it is a dense matrix. Therefore we have to find some inexpensive approximation $\hat{S} \approx S$ first. The choice of the approximation yields different preconditioners.

3.1.1 Least-Squares Commutator Preconditioner

The least-squares commutator (LSC) preconditioner proposed by Elman [4, 5] is based on the idea of approximate commutators similar to the pressure convection-diffusion (PCD) preconditioner [12]. The general algebraic idea of these methods is to find a matrix X for which

$$B^T X \approx F B^T \qquad (12)$$

and consequently

$$S = -B F^{-1} B^T \approx -B B^T X^{-1}, \qquad (13)$$

i.e., the inverse is moved so that it is no longer between the two rectangular matrices.

The construction of these preconditioners is based on the fact that F is a discretization of the operator

$$\mathcal{L} = \frac{1}{\Delta t} - \nu \Delta + (\mathbf{w} \cdot \nabla) \qquad (14)$$

defined on the discrete velocity space, where $\mathbf{w}$ is the approximation of the velocity computed in the previous Picard iteration. Suppose that an analogous operator is well defined also on the discrete pressure space and denote it by $\mathcal{L}_p$. It can be expected that the commutator with the gradient operator

$$\mathcal{E} = \mathcal{L} \nabla - \nabla \mathcal{L}_p \qquad (15)$$

is small for smooth $\mathbf{w}$ [5]. The discrete version of the commutator in terms of finite element matrices takes the form

$$E = (M_u^{-1} F)(M_u^{-1} B^T) - (M_u^{-1} B^T)(M_p^{-1} F_p), \qquad (16)$$

where M_u and M_p are the velocity and pressure mass matrix, respectively, and F_p is a discrete version of $\mathcal{L}_p$. Assume that E is also small, which means that

$$(M_u^{-1}F)(M_u^{-1}B^T) \approx (M_u^{-1}B^T)(M_p^{-1}F_p). \tag{17}$$

After several algebraic manipulations, this leads to the following approximation to the Schur complement

$$S = -BF^{-1}B^T \approx -BM_u^{-1}B^T F_p^{-1} M_p. \tag{18}$$

The LSC preconditioner avoids the explicit construction of F_p (unlike PCD) and defines the j-th column of F_p as a solution of the following weighted least-squares problem

$$\min \| [M_u^{-1}FM_u^{-1}B^T]_j - M_u^{-1}B^T M_p^{-1}[F_p]_j \|_{M_u}, \tag{19}$$

where $\|x\|_{M_u} = \sqrt{x^T M_u x}$ is a discrete analogue of the continuous L_2 norm on the velocity space. The vector $[F_p]_j$ is obtained by solving the normal equations

$$M_p^{-1}BM_u^{-1}B^T M_p^{-1}[F_p]_j = [M_p^{-1}BM_u^{-1}FM_u^{-1}B^T]_j, \tag{20}$$

which leads to the following definition of F_p:

$$F_p = M_p(BM_u^{-1}B^T)^{-1}(BM_u^{-1}FM_u^{-1}B^T). \tag{21}$$

Substituting this into (18) we get the following approximation of the Schur complement

$$\hat{S}_{LSC} = -(BM_u^{-1}B^T)(BM_u^{-1}FM_u^{-1}B^T)^{-1}(BM_u^{-1}B^T). \tag{22}$$

Since the inverse of the velocity mass matrix M_u^{-1} is dense, we replace M_u by its diagonal $\hat{M}_u$.

In Algorithm 2 we can see the individual steps of the LSC preconditioner application. It involves solving two subsystems with the matrix $A_L = B\hat{M}_u^{-1}B^T$, which is essentially a discrete Laplace operator. Thus, two Poisson-type solves for pressure and one velocity solve are needed.

Algorithm 2 LSC preconditioner

1: Solve $A_L z_p = r_p$, where $A_L = B\hat{M}_u^{-1}B^T$
2: Update $r_p = (B\hat{M}_u^{-1}F\hat{M}_u^{-1}B^T)z_p$
3: Solve $A_L z_p = -r_p$
4: Update $r_u = r_u - B^T z_p$
5: Solve $Fz_u = r_u$

3.1.2 Augmented Lagrangian Preconditioner

A different approach has been proposed by Benzi and Olshanskii [1]. The original system (5) is replaced with the equivalent system

$$\begin{bmatrix} F_\gamma & B^T \\ B & 0 \end{bmatrix} \begin{bmatrix} u \\ p \end{bmatrix} = \begin{bmatrix} f_\gamma \\ g \end{bmatrix}, \qquad (23)$$

where $F_\gamma = F + \gamma B^T W^{-1} B$, $f_\gamma = f + \gamma B^T W^{-1} g$, $\gamma > 0$ is a parameter and W is a positive definite matrix. The system (23) is then preconditioned with the block triangular preconditioner

$$P_{AL} = \begin{bmatrix} F_\gamma & B^T \\ 0 & \hat{S}_{AL} \end{bmatrix}, \qquad (24)$$

where the inverse of the Schur complement approximation is given by

$$\hat{S}_{AL}^{-1} := -\nu \tilde{M}_p^{-1} - \gamma W^{-1} \qquad (25)$$

and $\tilde{M}_p$ is a pressure mass matrix approximation, usually a diagonal matrix. The matrix W is often chosen to be equal to $\tilde{M}_p$.

Of course, the choice of the parameter γ is important. A large value would lead to small number of iterations of the preconditioned Krylov method, but for large γ the block F_γ becomes increasingly ill-conditioned and makes the solution of the subsystems expensive [17]. Hence, it is often set $\gamma \approx 1$.

The main difficulty of this approach is the choice of the approximate solver for the subsystems with F_γ. The additional term $\gamma B^T W^{-1} B$ makes the matrix less sparse compared to F and introduces a coupling between the velocity components which is not present in the discretization of the Picard linearized Navier–Stokes equations (without rotation or the periodic conditions mentioned in Sect. 2.2). The authors in [1] develop a multigrid method suitable for these subsystems.

Modified Version

One way to simplify the solution of the systems with F_γ is the modified version of AL preconditioner introduced in [2]. Let us denote the particular blocks of F_γ in two dimensions as follows

$$F_\gamma = \begin{bmatrix} A_{11} & A_{12} \\ A_{21} & A_{22} \end{bmatrix}. \qquad (26)$$

The modified approach suggests to replace this block by its upper block triangle

$$\tilde{F}_\gamma =: \begin{bmatrix} A_{11} & A_{12} \\ 0 & A_{22} \end{bmatrix}, \qquad (27)$$

such that instead of solving the whole system at once, we solve two smaller systems with the blocks A_{11} and A_{22}. These blocks can be interpreted as discrete anisotropic convection-diffusion operators, thus, applying $\tilde{F}_\gamma^{-1}$ requires solving two anisotropic convection-diffusion problems. The situation is similar in three dimensions, where we have to solve three subsystems.

3.2 SIMPLE-Type Preconditioners

SIMPLE (Semi-Implicit Method for Pressure Linked Equations) is an algorithm for numerical solution of the Navier–Stokes equations developed for finite volume and finite difference discretizations by Patankar and Spalding [16]. It is based on decoupling the system of equations and solving the velocity and pressure part separately. First, the velocity is solved from the momentum equations assuming that the pressure is known from the previous iteration. Then, the pressure and velocity are corrected in order to satisfy the discrete continuity equation.

This algorithm can be written in the form of block matrices and preconditioners for the systems arising from finite element discretizations can be derived. An overview of these preconditioners is given in [18]. The derivation is based on the block LU decomposition of the coefficient matrix in (5) and rewriting the system as

$$\begin{bmatrix} F & B^T \\ B & 0 \end{bmatrix} \begin{bmatrix} u \\ p \end{bmatrix} = \begin{bmatrix} F & 0 \\ B & S \end{bmatrix} \begin{bmatrix} I & F^{-1}B^T \\ 0 & I \end{bmatrix} \begin{bmatrix} u \\ p \end{bmatrix} = \begin{bmatrix} f \\ g \end{bmatrix}. \qquad (28)$$

The SIMPLE algorithm is obtained by using the approximation $F^{-1} \approx D^{-1} = \text{diag}(F)^{-1}$, introducing intermediate values u^*, p^* and corrections $\delta u, \delta p$ such that

$$u = u^* + \delta u, \quad p = p^* + \delta p, \qquad (29)$$

and solving the system in the following two steps:

$$\begin{bmatrix} F & 0 \\ B & \hat{S}_S \end{bmatrix} \begin{bmatrix} u^* \\ \delta p \end{bmatrix} = \begin{bmatrix} f \\ g \end{bmatrix} \qquad (30)$$

and

$$\begin{bmatrix} I & D^{-1}B^T \\ 0 & I \end{bmatrix} \begin{bmatrix} u \\ p \end{bmatrix} = \begin{bmatrix} u^* \\ \delta p \end{bmatrix}, \quad (31)$$

where $\hat{S}_S = -BD^{-1}B^T$. These steps are performed recursively, leading to the Algorithm 3, where the intermediate pressure p^* is estimated from the prior iterations. One iteration of the SIMPLE algorithm is used as preconditioner with $p^* = 0$.

Algorithm 3 SIMPLE algorithm

1: Solve $Fu^* = f - B^T p^*$
2: Solve $\hat{S}_S \delta p = g - Bu^*$
3: Update $u = u^* + \delta u$, where $\delta u = -D^{-1}B^T \delta p$
4: Update $p = p^* + \delta p$

There are several modifications of the algorithm. One of them is called SIMPLER, where p^* is obtained as a solution of the system

$$\hat{S}_S p^* = g - BD^{-1}((D - F)u^k + f), \quad (32)$$

where u^k is the velocity from the previous iteration in the original algorithm, but in case of preconditioner it is taken equal to zero.

Another variant is MSIMPLER algorithm, which is obtained from SIMPLER by replacing all occurrences of D by an approximation of the velocity mass matrix $\hat{M}_u$. The choice of $\hat{M}_u$ depends on the particular type of elements. For more details on this preconditioner, see [18].

4 Numerical Experiments

In this section we present results of some numerical experiments. The linear systems used in the experiments are obtained from an in-house isogeometric incompressible flow solver implemented in C++ within a framework of the G+Smo[1] library. G+Smo is an open-source object-oriented template C++ library, that implements the concept of IgA, based on abstract classes for geometry, discretization basis, assemblers, solvers etc. For more information about the library, see the documentation [14]. The linear algebra tools available in G+Smo are mostly inherited from the Eigen library [10].

---

[1] http://github.com/gismo, http://gismo.github.io.

For the discretization we use the isogeometric Taylor–Hood element, which means that the pressure basis is taken from the geometry and the velocity basis is obtained by p-refinement (degree elevation) of the pressure basis.

We implemented the selected preconditioners, namely LSC, AL, modified AL (MAL), SIMPLE, SIMPLER and MSIMPLER, also in the framework of the G+Smo library. For now, a direct solver (sparse LU decomposition from Eigen) is used for solving all subsystems in the preconditioners.

The numerical experiments were performed using a machine with the following parameters: Windows Server 2012, 2× Intel Xeon CP E5-2690 v2 @ 3.00GHz, 256 GB RAM.

In the experiments, we use full GMRES with various preconditioners to solve one linear system obtained after performing several Picard iterations in the steady case or several time steps in the unsteady case and we track the relative residual norm $||r||_2/||b||_2$, where b denotes the right-hand side, and the solution time in seconds.

4.1 Backward Step 2D

As a simple test example, we consider a 2D backward facing step domain consisting of three B-spline patches with conforming mesh, where the degrees of freedom on the interfaces are identified. The individual patches are described as B-splines of degree 3.

For comparison, we consider three meshes with different level of uniform refinement with 11,229, 42,005 and 162,309 degrees of freedom (DOFs). The coarsest mesh is shown in Fig. 3. The step height h as well as the inlet height is equal to 1. We prescribe a parabolic velocity profile with the maximum of 1 at the inlet boundary, zero velocity on the walls (upper and lower boundaries) and the "do-nothing" boundary condition at the outlet. For unsteady computations, the initial conditions are taken as the solution of the corresponding steady Stokes problem. The Reynolds number $Re = \frac{1}{\nu}$, where ν is the kinematic viscosity.

In Table 1 we present the number of iterations and the computational time in seconds (in parentheses) needed to reach the relative residual norm smaller than 10^{-10} for the individual preconditioners in the steady (top) and unsteady (bottom) case. The computational time is reported in two separate parts, the time of the preconditioner setup involving the factorization of the subsystem matrices and the

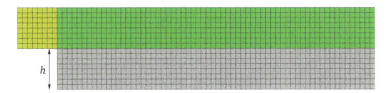

Fig. 3 The computational mesh with 11,229 DOFs

Table 1 Number of iterations and computational time in seconds (in parentheses) needed to fulfill $||r||_2/||b||_2 < 10^{-10}$, backward step steady (top) and unsteady (bottom) case for $Re = 100$

Steady	Mesh 1	Mesh 2	Mesh 3
LSC	32 (0.60 + 0.55)	40 (4.34 + 3.55)	55 (32.8 + 22.5)
SIMPLE	171 (0.47 + 3.19)	> 200	> 200
SIMPLER	28 (**0.49 + 0.51**)	28 (**3.87 + 2.59**)	37 (**30.4 + 15.02**)
MSIMPLER	32 (0.48 + 0.58)	40 (3.87 + 3.70)	54 (30.4 + 22.2)
AL	15 (10.8 + 1.01)	10 (232 + 5.70)	8 (4592 + 33.8)
MAL	> 200	> 200	> 200
Unsteady	Mesh 1	Mesh 2	Mesh 3
LSC	11 (0.70 + 0.22)	11 (4.95 + 1.12)	11 (33.1 + 4.71)
SIMPLE	34 (0.57 + 0.62)	29 (4.42 + 2.53)	22 (31.0 + 7.67)
SIMPLER	12 (**0.59 + 0.26**)	11 (4.47 + 1.16)	11 (**31.5 + 4.91**)
MSIMPLER	12 (**0.59 + 0.26**)	11 (**4.46 + 1.16**)	12 (31.6 + 5.33)
AL	47 (10.6 + 3.02)	47 (215 + 22.2)	47 (4063 + 168)
MAL	49 (3.46 + 2.07)	51 (80.4 + 14.6)	62 (1482 + 130)

time of the subsequent GMRES iterations. The lowest total time for a given problem is displayed in bold. The value $\gamma = 1$ was used for the AL and MAL preconditioners in these experiments and 200 was the maximum number of iterations. In the steady case, the number of iterations for LSC and SIMPLE-type preconditioners increases for finer meshes. On the contrary, the number of iterations for AL decreases, but this preconditioner is obviously very expensive with the direct solver for the subsystems, hence it is really necessary to use some efficient iterative method to solve them. The convergence of the modified AL is very slow, probably because $\gamma = 1$ is far from the optimal value of γ for MAL in this case. In the unsteady case, the convergence is generally faster than in the steady case for most preconditioners and most of them (except MAL) seem independent of the mesh size.

We note that in real applications, especially three-dimensional problems, the subsystems cannot be solved by direct solution methods. We refer to [13] and [18] where comparable problems are solved (using finite volume and finite element method). It appears that the solution of the subsystems can be approximated by 1 or 2 iterations of a MG V-cycle, or a solution with PCG or preconditioned Bi-CGSTAB with a moderate accuracy (reduction of the relative residual by a factor 10 or 100). Since the preconditioner is used to find a suitable search direction for the outer Krylov method, it appears that it is not very sensitive to the accuracy of the inner solves. If MG can be used, the total solver becomes scalable.

In the next experiment, we consider a steady flow with $\nu = 0.01, 0.009, \ldots,$ 0.003 and an unsteady flow with $\nu = 10^{-2}, 10^{-3}, 10^{-4}$ in the backward step domain with mesh 2 to study the dependence on the Reynolds number. Here, the value of γ for MAL was chosen as the "optimal" value from the interval $[0.1, 2.5]$ (found experimentally with step 0.1). Figure 4 shows the evolution of the relative residual norm during 100 iterations of GMRES for the individual preconditioners in the steady case. We can see that the convergence slows down after several first iterations

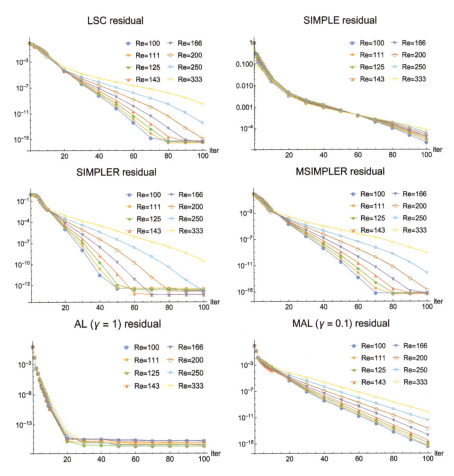

Fig. 4 Residual evolution for the individual preconditioners, a linear system from the steady Navier–Stokes with Reynolds number Re varying between 100 and 333, backward step with mesh 2

for increasing Reynolds number for all preconditioners except AL. A comparison of the preconditioners for the lowest and highest Reynolds numbers $Re = 100$ and $Re = 333$ is given in the left and right picture of Fig. 5, respectively. The convergence of AL is significantly faster than of all the other preconditioners.

Figure 6 shows the evolution of the relative residual norm in the unsteady case. The convergence of all preconditioners is almost independent of the Reynolds number, except for SIMPLE, for which it is slightly dependent. AL does not show superior convergence behavior anymore in the unsteady case.

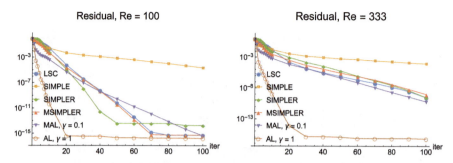

Fig. 5 Residual evolution for various preconditioners, a linear system from steady Navier–Stokes, backward step with mesh 2, $Re = 100$ (left) and $Re = 333$ (right)

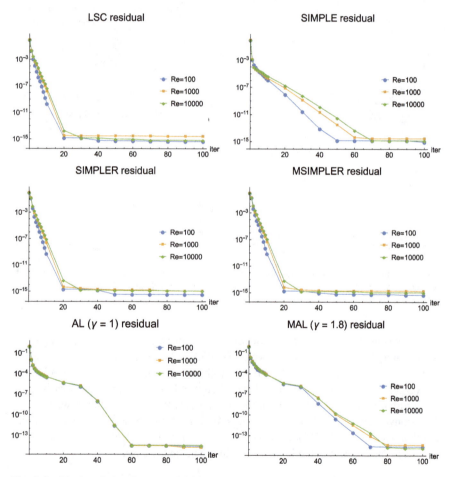

Fig. 6 Residual evolution for the individual preconditioners, a linear system from the unsteady Navier–Stokes with Reynolds number Re varying between 100 and 10000, backward step with mesh 2

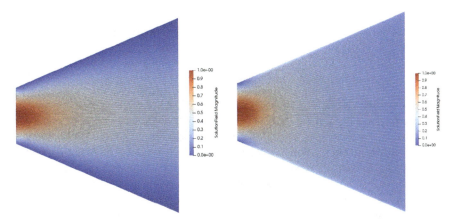

Fig. 7 Velocity with streamlines in a trapezoidal domain where the lower and upper boundaries are solid walls (left) or periodic sides (right)

4.2 Periodic Domain 2D

In order to test the effect of periodic boundary conditions on the convergence of the iterative solvers, consider a simple (artificial) test example in 2D – a flow in a trapezoidal domain shown in Fig. 7. The left and right boundary is the inflow and outflow boundary, respectively, and the lower and upper boundaries are either solid walls or periodic sides. The pictures display the velocity solution of an unsteady Navier-Stokes problem with viscosity $\nu = 10^{-2}$ after 20 time steps with $\Delta t = 10^{-2}$. The geometry is described as a B-spline of degree 3, hence the degree of the basis functions is 3 for pressure and 4 for velocity. The experiments were done for three meshes, a uniformly refined mesh with around 10,000 DOFs and two meshes with different levels of local refinement near the upper and lower boundary with around 11,000 and 12,000 DOFs. The maximum aspect ratio of the meshes is approximately 4, 16 and 64, respectively. Such local refinement is of interest, because in the real computations in the water turbine we simulate turbulent flow using the RANS equations with a turbulence model and therefore we need to refine near the vanes/blades (i.e. the periodic sides).

In the periodic case, where the block F is no longer block diagonal (see (8)), the subsystems with F cannot be split into several smaller subsystems corresponding to the velocity components. In the 2D example, the system structure is as in (8) without the first block row and column. For experimental purposes, we implemented several ways to solve these subsystems (exactly or approximately):

- `Fdiag`

    We neglect all off-diagonal blocks and split the system into d subsystems with the diagonal blocks.

- Fmod

  We replace F by its upper block triangle, similarly to the modified AL approach.
- Fwhole

  We solve the whole system $Fz_u = r_u$.

We do not apply these variants to the AL preconditioner, since it requires solution of systems with the matrix F_γ instead of F, which is already not block diagonal and the whole system has to be solved even without periodic conditions. We remind that MAL is a modification of AL replacing F_γ with its upper block triangle (in both periodic and nonperiodic case).

In Tables 2 and 3 we present the number of iterations and the computational time needed to reach the relative residual norm smaller than 10^{-10} for the individual preconditioners in the nonperiodic and periodic case, respectively. Again, $\gamma = 1$ was used for the AL and MAL preconditioners in these experiments.

It can be seen from Table 2 that the convergence of most of the preconditioners (except SIMPLER and AL) is dependent on the local refinement, i.e. the aspect ratio of the mesh, in the nonperiodic case.

In the periodic case (Table 3), we compare the approaches to the subsystems with block F described above. For the uniformly refined mesh, there are only small differences in the number of iterations for different approaches, hence, it seems to be sufficient to use Fdiag in this case. However, for the locally refined meshes the differences become more significant. Further, LSC, SIMPLER and MSIMPLER give similar numbers of iterations for all meshes with Fwhole (i.e. if the subsystem with F is solved exactly), but the dependence on the aspect ratio with the approximations of F is more significant for LSC than for SIMPLER and MSIMPLER. SIMPLER is no longer independent of the aspect ratio even with Fwhole. The AL preconditioner gives exactly the same number of iterations as in the nonperiodic case and stays independent of the aspect ratio, but it is very expensive. The modified version of AL needs approximately twice as many iterations and half the computational time compared to AL with the same γ in the case of uniform refinement, but its convergence is dependent on the aspect ratio and the dependence is much stronger in the periodic case. Note that $\gamma = 1$ is not an

Table 2 Number of iterations and computational time in seconds (in parentheses) needed to fulfill $||r||_2/||b||_2 < 10^{-10}$, nonperiodic trapezoidal domain

Precond.	Uniform	loc1	loc2
LSC	11 (0.47 + 0.14)	13 (0.56 + 0.20)	21 (0.66 + 0.37)
SIMPLE	27 (0.36 + 0.32)	28 (0.43 + 0.39)	34 (0.51 + 0.57)
SIMPLER	11 (**0.37 + 0.16**)	11 (**0.45 + 0.19**)	11 (**0.53 + 0.22**)
MSIMPLER	12 (0.37 + 0.17)	14 (0.45 + 0.23)	22 (0.53 + 0.43)
AL	22 (7.03 + 1.09)	22 (8.83 + 1.29)	22 (9.80 + 1.43)
MAL	41 (3.05 + 1.46)	53 (3.61 + 2.18)	63 (3.96 + 2.89)

Bold values indicate the lowest total computational time

Table 3 Number of iterations and computational time in seconds (in parentheses) needed to fulfill $||r||_2/||b||_2 < 10^{-10}$, periodic trapezoidal domain

Precond.	Uniform	loc1	loc2
LSC			
Fdiag	18 (0.73 + 0.26)	33 (0.95 + 0.57)	70 (1.20 + 1.48)
Fmod	15 (0.73 + 0.22)	24 (0.95 + 0.42)	50 (1.21 + 1.05)
Fwhole	11 (1.26 + 0.22)	13 (1.64 + 0.30)	17 (1.99 + 0.45)
SIMPLE			
Fdiag	31 (0.61 + 0.42)	38 (0.81 + 0.63)	61 (1.05 + 1.21)
Fmod	29 (0.61 + 0.41)	36 (0.81 + 0.61)	54 (1.05 + 1.09)
Fwhole	28 (1.13 + 0.51)	31 (1.50 + 0.67)	42 (1.84 + 1.05)
SIMPLER			
Fdiag	13 (0.62 + 0.20)	22 (0.82 + 0.41)	44 (1.06 + 0.97)
Fmod	12 (**0.62 + 0.19**)	17 (**0.83 + 0.32**)	34 (**1.06 + 0.76**)
Fwhole	11 (1.15 + 0.23)	16 (1.52 + 0.38)	26 (1.85 + 0.71)
MSIMPLER			
Fdiag	14 (0.62 + 0.22)	22 (0.82 + 0.41)	45 (1.06 + 0.99)
Fmod	13 (0.62 + 0.21)	18 (0.82 + 0.34)	35 (1.06 + 0.78)
Fwhole	12 (1.15 + 0.25)	13 (1.51 + 0.32)	19 (1.85 + 0.53)
AL	22 (6.78 + 1.29)	22 (9.04 + 1.43)	22 (12.0 + 1.68)
MAL	46 (2.16 + 1.70)	92 (2.97 + 4.03)	195 (4.08 + 10.6)

Bold values indicate the lowest total computational time

optimal value for MAL in this case. We can get a bit faster convergence for a better choice of γ, but the aspect ratio dependence is similar.

5 Conclusions

In this paper we studied convergence behavior of selected block triangular preconditioners (LSC, AL, MAL) and SIMPLE-type preconditioners (SIMPLE, SIMPLER, MSIMPLER) for linear systems obtained from IgA discretization of steady and unsteady incompressible Navier–Stokes equations. All subsystems in the preconditioners were solved with a direct solver.

We investigated the dependence on the mesh size and the Reynolds number for a simple 2D backward facing step example. The experiments show, that most of the preconditioners are dependent on both, the mesh size and the Reynolds number, in the steady case, but independent or almost independent in the unsteady case. Further, it seems that AL outperforms the other preconditioners in terms of number of iterations in the steady case, but it is much more expensive, at least if the subsystems are solved directly. The convergence of LSC and SIMPLE-type preconditioners is generally faster for the unsteady problem than for the steady problem. Such behavior is expectable due to the nature of the systems. The main block diagonal of the matrix

F in the linear system arising from the unsteady case contains the velocity mass matrix multiplied by $1/\Delta t$. Thus, the mass matrix becomes a dominant part of the system matrix, which makes the system easier to solve.

Then we considered a simple 2D domain to simulate some of the typical features in the water turbines - periodic boundary conditions on nonparallel sides and locally refined meshes along them. We tested several simplifications of the solution of the system with block F, which is no longer block diagonal in the periodic case. In general, all tested preconditioners except AL are dependent on the aspect ratio. According to the experiments, we can neglect the off-diagonal blocks of F in the periodic case for low aspect ratio, but it seems necessary to solve the system with the full block F for higher aspect ratios. This issue is a topic for further research.

Another interesting topic for future research is the dependence of the convergence on the order of continuity of the IgA solution. We also plan to focus on iterative solution of the subsystems in the preconditioners in the future.

Acknowledgments This work has been supported by the European Union's Horizon 2020 research and innovation programme under grant agreement No. 678727 and by the project LO1506 of the Czech Ministry of Education, Youth and Sports under the program NPU I.

References

1. M. Benzi and M. A. Olshanskii. An augmented Lagrangian-based approach to the Oseen problem. *SIAM J. Sci. Comput.*, 28(6):2095–2113, 2006.
2. M. Benzi, M. A. Olshanskii, and Z. Wang. Modified augmented Lagrangian preconditioners for the incompressible Navier–Stokes equations. *Int. J. Numer. Meth. Fluids*, 66(4):486–508, 2011.
3. A. Buffa, C. de Falco, and G. Sangalli. Isogeometric analysis: stable elements for the 2D Stokes equation. *International Journal for Numerical Methods in Fluids*, 65(11–12):1407–1422, 2011.
4. H. Elman. Preconditioning for the steady-state Navier–Stokes equations with low viscosity. *SIAM J. Sci. Comput.*, 20(4):1299–1316, 1999.
5. H. Elman, V. E. Howle, J. Shadid, R. Shuttleworth, and R. Tuminaro. Block preconditioners based on approximate commutators. *SIAM J. Sci. Comput.*, 27(5):1651–1668, 2006.
6. H. Elman, V. E. Howle, J. Shadid, D. Silvester, and R. Tuminaro. Least squares preconditioners for stabilized discretizations of the Navier–Stokes equations. *SIAM J. Sci. Comput.*, 30(1):290–311, 2007.
7. J. A. Evans and T. J. R. Hughes. Isogeometric divergence-conforming B-splines for the steady Navier–Stokes equations. *Math. Models Methods Appl. Sci.*, 23(8):1421–1478, 2013.
8. J. A. Evans and T. J. R. Hughes. Isogeometric divergence-conforming B-splines for the unsteady Navier–Stokes equations. *J. Comput. Phys.*, 241:141–167, 2013.
9. J. Fouchet-Incaux. Artificial boundaries and formulations for the incompressible Navier–Stokes equations. Applications to air and blood flows. *SeMA Journal*, 64:1–40, 2014.
10. G. Guennebaud, B. Jacob, et al. Eigen v3. http://eigen.tuxfamily.org, 2010.
11. J. G. Heywood, R. Rannacher, and S. Turek. Artificial boundaries and flux and pressure conditions for the incompressible Navier–Stokes equations. *Int. J. Numer. Meth. Fluids*, 22:325–352, 1996.
12. D. Kay, D. Loghin, and A. Wathen. A preconditioner for the steady-state Navier–Stokes equations. *SIAM J. Sci. Comput.*, 24(1):237–256, 2002.

13. C. M. Klaij and C. Vuik. SIMPLE-type preconditioners for cell-centered, colocated finite volume discretization of incompressible Reynolds-averaged Navier-Stokes equations. *Int. J. Numer. Meth. Fluids*, 71:830–849, 2013.
14. A. Mantzaflaris and others (see website). G+Smo (Geometry plus Simulation modules) v0.8.1. http://github.com/gismo, 2018.
15. M. F. Murphy, G. H. Golub, and A. J. Wathen. A note on preconditioning for indefinite linear systems. *SIAM J. Sci. Comput.*, 21(6):1969–1972, 2000.
16. S. V. Patankar and D. B. Spalding. A calculation procedure for heat, mass and momentum transfer in three-dimensional parabolic flows. *Int. J. Heat and Mass Transfer*, 15:1787–1805, 1972.
17. A. Segal, M. ur Rehman, and C. Vuik. Preconditioners for incompressible Navier–Stokes solvers. *Numer. Math. Theor. Meth. Appl.*, 3(3):245–275, 2010.
18. M. ur Rehman, C. Vuik, and G. Segal. SIMPLE-type preconditioners for the Oseen problem. *Int. J. Numer. Meth. Fluids*, 61(4):432–452, 2009.

Solving 2D Heat Transfer Problems with the Aid of a BEM-Isogeometric Solver

Konstantinos Kostas, Yeraly Kalel, and Azat Amiralin

Abstract In this chapter, we present a computational framework that performs design optimization via seamless integration of geometric modeling and computational analysis on the basis of the isogeometric analysis concept. The framework comprises a parametric geometric modeler, a Boundary Element Method empowered by the IsoGeometric Analysis approach (IGABEM) and a set of gradient-based and meta-heuristic, evolutionary optimization algorithms that drive the design optimization. We demonstrate our approach in the case of a 2D steady-state heat conduction problem across a periodic interface separating two conducting and conforming media, where our goal is to either generate the interface that maximizes heat transfer or identify the geometry able to produce a given heat transfer rate and/or temperature distribution along the unknown separating geometry. The parametric modeler generates instances of the separating interface geometry, accurately represented via a NURBS representation, which are assessed by our IGABEM solver that numerically solves the system of Boundary Integral Equations (BIE) arising in the context of the 2D steady-state heat conduction problem across the periodic interface of the bilayered structure mentioned above. The efficiency of our approach and the performance of the developed computational framework is demonstrated for both design optimization and inverse design problems.

1 Introduction

This chapter describes the development of a robust design optimization framework that combines the work carried out for the investigation of optimum corrugations, and optimal design of high conductivity inserts and fins carried out by Leontiou et al. [18] and Fyrillas et al. [9], and the introduction of an IsoGeometric Boundary Element Method (IGABEM), by Kostas et al. [17], which efficiently handles the

K. Kostas (✉) · Y. Kalel · A. Amiralin
School of Engineering, Nazarbayev University, Astana, Kazakhstan
e-mail: konstantinos.kostas@nu.edu.kz

numerical solution of the governing partial differential equations for the case of an arbitrary smooth, periodic interface that separates two infinite-length slabs in conduct with different conductivity coefficients.

The idea of the seamless integration between Computer-Aided Design (CAD) and Computer-Aided Analysis (CAA) via the use of IsoGeometric Analysis (IGA), as described in Hughes et al. [13] and Cottrell et al. [8], forms the basis of our design optimization framework. The IGA approach directly employs analysis-suitable geometric models from the CAD representation for the conducted analysis and this can be efficiently exploited for the case of shape optimization and inverse design, which are the major objectives in our work. Shape optimization in the context of IGA has been already presented in various works, as e.g., Wall et al. [39], Nagy et al. [24], Nguyen et al. [25], Nørtoft and Gravesen [26], Qian and Sigmund [32], Cho and Ha [6], Lian et al. [20], Qian [31], Li and Qian [19], Sun et al. [38] where NURBS control points and/or weights have been generally used as design variables to control the boundary shape. Furthermore, Kostas et al. [15, 16], Kaklis et al. [14] and Ginnis et al. [10] have already presented higher-level geometric parametric models that employ high-level parameters that encapsulate meaningful quantities, from an engineering and/or design point of view, for ship hulls and hydrofoils optimum designs. Although the parametric models employed in this study do not exhibit the complexity of a ship-hull surface model, we have tried to incorporate all the mechanisms that guarantee robustness in the generation of instances (elimination of invalid shapes) and minimization of design variables to the extent possible.

The design problem in our case involves the case of a 2D steady-state heat conduction problem across a periodic interface separating two conducting and conforming media, where our goal is to either generate the interface that maximizes heat transfer or identify the geometry able to produce a given heat transfer rate and/or temperature/heat-flux distribution along the unknown separating geometry. These sets of problems are numerically solved using the IGABEM approach presented in [17]. The IGA concept was initially introduced in the context of the Finite Element Method by Hughes et al, as we have already mentioned above. This concept was later extended to the Boundary Element Method (BEM) by various authors, see, e.g., Politis et al. [28], Simpson et al. [35], Scott et al. [33], Belibassakis et al. [4], Peake et al. [27], Simpson et al. [36], Ginnis et al. [11], Kaklis et al. [14]. Recently, An et al. [2] have also employed the isogeometric-BEM (IGA-BEM) concept for 2D steady heat transfer analysis with very good results. In our work, the IGABEM approach is based on a NURBS representation of the interface curve and employs the very same basis of the geometry for representing the temperature (T) and its normal derivative ($\frac{\partial T}{\partial \mathbf{n}}$) on the interface. The Boundary Integral Equations are numerically solved by collocating at the knotvector's Greville abscissae for the interface's parametric representation. As discussed in [17], this IGABEM approach exhibits superior convergence rate and achieves the same level of accuracy with a significantly smaller number of degrees of freedom and considerably less computational time when compared to a classical low-order BEM implementation.

1.1 Using the Design Optimization Framework

In this chapter, we present the use of the developed design optimization framework in providing answers or, more precisely, demonstrating the way of seeking answers, to a family of questions related to temperature and heat transfer between two infinite, conducting and conforming media (materials) with different conductivity coefficients in contact over a periodic separating interface with isothermal conditions on both, lower and upper, boundaries:

- What is the shape of the interface that maximizes heat transfer under given media-area and/or interface-length constraints?
- If the heat transfer value is given, which is(are) the shape(s) of the interface(s) that results in that value?
- If we prescribe a heat transfer value larger than what can be reached, what is the answer to the problem above?
- If the temperature distribution (or the heat flux) along the separating interface is prescribed or measured, can we compute the shape of the interface itself?
- and other similar questions.

1.2 Chapter Outline

We begin the presentation of our work by discussing the problem formulation (see Sect. 2.1) and describing, in detail, the IGABEM approach for solving the heat transfer problem; see Sect. 2.2.

Section 3 is devoted to the presentation of the parametric models we use for the representation of a periodic free-form interface employed in our design framework. The two families of parametric models are discussed in Sects. 3.1 and 3.2, respectively.

The next section (Sect. 4) comprises the major part of this chapter and begins with a discussion of the optimization algorithms that have been used and tested in the context of our design framework. This discussion is followed by two different application cases and a series of examples regarding the forward design case (Sect. 4.1) and the inverse design case (Sect. 4.2). Both cases are handled using the same design optimization framework and the examples included range from the determination of optimum fins to identification of the shape that produces a given temperature distribution. The last section summarizes the presented results and refers to some possible future enhancements and extensions.

2 Heat Transfer Problem Formulation

In all shape optimization and inverse design cases presented in this chapter, we consider a 2D steady-state heat conduction problem. The heat transfer problem is investigated across a periodic interface that separates two conducting and conforming media. The separating interface S is a planar, generally continuous, free-form curve, which is periodic with a period L. For the boundaries of both media that are not in conduct, we consider isothermal conditions as it is depicted in Fig. 1.

2.1 Continuous Formulation

As our separating interface exhibits longitudinal periodicity, it suffices to examine our problem within one period of length L. The temperature fields in both media will satisfy the Laplace equation, i.e. $\Delta T_1 = 0$ and $\Delta T_2 = 0$, and additionally will need to satisfy the isothermal conditions, i.e. $T_1(0, x) = c_1$ and $T_2(H, x) = c_2$ where $H = h_1(x) + h_2(x)$ is the total height of the bilayered structure, the periodic conditions at the vertical walls $x = 0$ and $x = L$ and finally, the matching conditions (temperature and flux continuity) along the common boundary S.

Without loss of generality we may transform both temperature fields by setting $\tilde{T}_1 = T_1 - c_1$ and $\tilde{T}_2 = T_2 - c_2$. Further to this, we may nondimensionalize lengths (H, h_1, h_2, L) with the aid of period's length L. Finally, for reasons of notational simplification, let us denote from this point onwards $\tilde{T}_1$ as T_1 and $\tilde{T}_2$ as T_2. In view of these definitions and assumptions, we can state our Boundary Value Problem

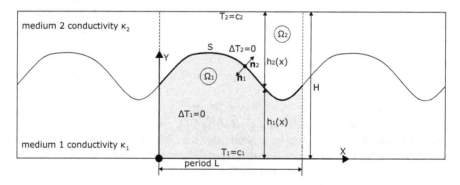

Fig. 1 L-periodic interface between two conducting and conforming media

(BVP) formulation as follows:

$$\Delta T_1(\mathbf{p}_1) = 0, \quad \mathbf{p}_1 = (x, y) \in \Omega_1 \quad \text{and} \quad \Delta T_2(\mathbf{p}_2) = 0, \quad \mathbf{p}_2 = (x, y) \in \Omega_2, \tag{1}$$

with the following boundary conditions:

1. **Isothermal conditions**:

$$T_1(x, 0) = 0, \quad T_2(x, H) = 0, \quad x \in [0, 1] \tag{2a}$$

2. **Periodic conditions**:

$$T_1(0, y) = T_1(1, y), \quad \frac{\partial T_1(0, y)}{\partial x} = \frac{\partial T_1(1, y)}{\partial x}, \quad y \in [0, h_1(0) = h_1(1)] \tag{2b}$$

$$T_2(0, y) = T_2(1, y), \quad \frac{\partial T_2(0, y)}{\partial x} = \frac{\partial T_2(1, y)}{\partial x}, \quad y \in [h_1(0) = h_1(1), H] \tag{2c}$$

3. **Interface conditions**:

$$T_1(\mathbf{p}) + c_1 = T_2(\mathbf{p}) + c_2, \quad \mathbf{p} \in S \tag{2d}$$

$$\kappa_1 \frac{\partial T_1(\mathbf{p})}{\partial \mathbf{n}_1} = -\kappa_2 \frac{\partial T_2(\mathbf{p})}{\partial \mathbf{n}_2}, \quad \mathbf{p} \in S, \tag{2e}$$

where $\mathbf{n}_1, \mathbf{n}_2$ correspond to interface's normal vectors as they are depicted in Fig. 1, respectively. If we further assume, without loss of generality, that $c_1 = 1$ and $c_2 = 0$, Eq. 2d may be written as

$$T_1(\mathbf{p}) + 1 = T_2(\mathbf{p}), \quad \mathbf{p} \in S. \tag{2f}$$

We choose to solve this BVP problem with the Boundary Integral Equation (BIE) method which requires the use of periodic Green's function for the Laplace equations in Eq. 1. As it is described by Pozrikidis in [30], the Green's functions for the two domains, which satisfy the isothermal and periodic conditions, are expressed as follows:

$$G_1(\mathbf{p}, \mathbf{p}_0) = \frac{1}{4\pi} \ln\left(\cosh(2\pi(y - y_0)) - \cos(2\pi(x - x_0))\right) -$$

$$\frac{1}{4\pi} \ln\left(\cosh(2\pi(y + y_0)) - \cos(2\pi(x - x_0))\right), \tag{3}$$

$$\mathbf{p} = (x, y), \mathbf{p}_0 = (x_0, y_0) \in \Omega_1$$

$$G_2(\mathbf{p}, \mathbf{p}_0) = \frac{1}{4\pi} \ln\left(\cosh(2\pi(y - y_0)) - \cos(2\pi(x - x_0))\right) -$$
$$\frac{1}{4\pi} \ln\left(\cosh(2\pi(y + y_0 - 2H)) - \cos(2\pi(x - x_0))\right), \quad (4)$$
$$\mathbf{p} = (x, y), \mathbf{p}_0 = (x_0, y_0) \in \Omega_2$$

The application of Green's second identity for the temperature field and corresponding Green's functions in each of the problem's domains lead to the following Fredholm-type equations of the second kind:

$$\frac{T_1(\mathbf{p}_0)}{2} = \int_S G_1(\mathbf{p}, \mathbf{p}_0) \frac{\partial T_1(\mathbf{p})}{\partial \mathbf{n}_1(\mathbf{p})} d\ell(\mathbf{p}) - \int_S T_1(\mathbf{p}) \frac{\partial G_1(\mathbf{p}, \mathbf{p}_0)}{\partial \mathbf{n}_1(\mathbf{p})} d\ell(\mathbf{p}), \quad \mathbf{p}, \mathbf{p}_0 \in S, \quad (5a)$$

$$\frac{T_2(\mathbf{p}_0)}{2} = \int_S G_2(\mathbf{p}, \mathbf{p}_0) \frac{\partial T_2(\mathbf{p})}{\partial \mathbf{n}_2(\mathbf{p})} d\ell(\mathbf{p}) - \int_S T_2(\mathbf{p}) \frac{\partial G_2(\mathbf{p}, \mathbf{p}_0)}{\partial \mathbf{n}_2(\mathbf{p})} d\ell(\mathbf{p}), \quad \mathbf{p}, \mathbf{p}_0 \in S \Rightarrow$$

$$\frac{T_1(\mathbf{p}_0) + 1}{2} = -\int_S G_2(\mathbf{p}, \mathbf{p}_0) \frac{\kappa_1 \partial T_1(\mathbf{p})}{\kappa_2 \partial \mathbf{n}_1(\mathbf{p})} d\ell(\mathbf{p}) + \int_S (T_1(\mathbf{p}) + 1) \frac{\partial G_2(\mathbf{p}, \mathbf{p}_0)}{\partial \mathbf{n}_1(\mathbf{p})} d\ell(\mathbf{p}). \quad (5b)$$

The second equation can be easily derived with the aid of the interface conditions (Eqs. 2e, 2f) and the obvious observation that $\mathbf{n}_1 = -\mathbf{n}_2$.

Equations 5a and 5b constitute the system of equations we need to solve with respect to $T_1(\mathbf{p})$ and $\frac{\partial T_1(\mathbf{p})}{\partial \mathbf{n}_1(\mathbf{p})}$, $\mathbf{p} \in S$. Obviously, if we know the heat flux along one period of the separating interface, the corresponding dimensionless heat transfer across it can be easily computed as:

$$h_T = \int_S \frac{\partial T_1(\mathbf{p})}{\partial \mathbf{n}_1(\mathbf{p})} d\ell(\mathbf{p}). \quad (6)$$

2.2 IGABEM Formulation

The essence of the IGA approach boils down to expressing the unknown field quantities, temperature and normal flux in our case, using the exact same basis that is employed in the representation of the geometry, i.e., one period of the separating interface in our problem. In our case, we follow an isogeometric boundary element method approach (IGABEM) to solve the BIE system derived in Sect. 2.1 and, for all cases presented in this chapter, we assume that one period of the separating interface S is accurately represented by a regular parametric NURBS curve:

$$S(t) = \hat{\mathbf{p}}(t) = \sum_{i=0}^{n} \hat{\mathbf{b}}_i N_{i,k}(t), \ t \in [0, 1], \quad \text{in homogeneous coordinates or}$$

$$S(t) = \mathbf{p}(t) = (p_x(t), p_y(t)) = \sum_{i=0}^{n} \mathbf{b}_i R_{i,k}(t), \ t \in [0, 1], \text{ in Cartesian coordinates}, \quad (7)$$

where $\{N_{i,k}\}$, $i = 0,\ldots,n$ are the kth order B-spline basis functions defined over a knotvector $\mathscr{T} = \{t_0, t_1, \ldots, t_{n+k}\}$, $R_{i,k}(t) = \frac{w_i N_{i,k}(t)}{\sum_{i=0}^{n} w_i N_{i,k}(t)}$ are the kth order NURBS basis functions defined over the same knotvector, $\{\hat{\mathbf{b}}_i\}$, $i = 0,\ldots,n$ are curve's control points in homogeneous coordinates and finally $\{\mathbf{b}_i = (\frac{\hat{b}_{ix}}{w_i}, \frac{\hat{b}_{iy}}{w_i})\}$, $i = 0,\ldots,n$ are the control points in Cartesian coordinates with w_i being the corresponding control point weight. Using this representation (Eq. 7) for the separating interface S, we may obviously rewrite the system of BIE equations (Eqs. 5a, 5b) as follows:

$$\int_0^1 G_1(t,\tau) \frac{\partial T_1(t)}{\partial \mathbf{n}_1(t)} \|\dot{\mathbf{p}}(t)\| dt - \int_0^1 T_1(t) \frac{\partial G_1(t,\tau)}{\partial \mathbf{n}_1(t)} \|\dot{\mathbf{p}}(t)\| dt - \frac{T_1(\tau)}{2} = 0, \quad (8)$$

$$-\frac{\kappa_1}{\kappa_2} \int_0^1 G_2(t,\tau) \frac{\partial T_1(t)}{\partial \mathbf{n}_1(t)} \|\dot{\mathbf{p}}(t)\| dt + \int_0^1 T_1(t) \frac{\partial G_2(t,\tau)}{\partial \mathbf{n}_1(t)} \|\dot{\mathbf{p}}(t)\| dt - \frac{T_1(\tau)}{2} =$$

$$= \frac{1}{2} - \int_0^1 \frac{\partial G_2(t,\tau)}{\partial \mathbf{n}_1(t)} \|\dot{\mathbf{p}}(t)\| dt, \quad (9)$$

where $t,\tau \in [0,1]$ and obviously $T_1(t) = T_1(\mathbf{p}(t))$, $G(t,\tau) = G(\mathbf{p}(t),\mathbf{p}(\tau))$. The next step comprises the representation of the unknown quantities, $T_1(t)$ and $\frac{\partial T_1(t)}{\partial \mathbf{n}_1(t)}$ using the NURBS basis functions employed in Eq. 7. Additionally, we may allow for further refinement of this spline space ($\mathscr{S}^k(\mathscr{T})$) by knot insertions in $\mathscr{T}$. Hence, we may generally consider the insertion of m knots that will obviously preserve the shape of our interface S and at the same produce a refined spline space $\mathscr{S}^k(\mathscr{T}) \subset \mathscr{S}^k(\mathscr{T}^{(m)})$. Therefore, assuming that

$$T_1(t) = \sum_{i=0}^{n+m} T_{1,i} R_{i,k}^{(m)}(t), \quad (10)$$

$$F_1(t) = \frac{\partial T_1(t)}{\partial \mathbf{n}_1(t)} = \sum_{i=0}^{n+m} F_{1,i} R_{i,k}^{(m)}(t), \quad (11)$$

where $t \in [0,1]$ and $R_{i,k}^{(0)}(t) = R_{i,k}(t)$, we may rewrite our system of BIEs as follows:

$$\int_0^1 G_1(t,\tau) \sum_{i=0}^{n+m} F_{1,i} R_{i,k}^{(m)}(t) \|\dot{\mathbf{p}}(t)\| dt - \int_0^1 \sum_{i=0}^{n+m} T_{1,i} R_{i,k}^{(m)}(t) \frac{\partial G_1(t,\tau)}{\partial \mathbf{n}_1(t)} \|\dot{\mathbf{p}}(t)\| dt -$$

$$\frac{\sum_{i=0}^{n+m} T_{1,i} R_{i,k}^{(m)}(t)}{2} = 0, \quad (12a)$$

$$-\frac{\kappa_1}{\kappa_2} \int_0^1 G_2(t,\tau) \sum_{i=0}^{n+m} F_{1,i} R_{i,k}^{(m)}(t) \|\dot{\mathbf{p}}(t)\| dt + \int_0^1 \sum_{i=0}^{n+m} T_{1,i} R_{i,k}^{(m)}(t) \frac{\partial G_2(t,\tau)}{\partial \mathbf{n}_1(t)} \|\dot{\mathbf{p}}(t)\| dt -$$

$$\frac{\sum_{i=0}^{n+m} T_{1,i} R_{i,k}^{(m)}(t)}{2} = \frac{1}{2} - \int_0^1 \frac{\partial G_2(t,\tau)}{\partial \mathbf{n}_1(t)} \|\dot{\mathbf{p}}(t)\| dt. \quad (12b)$$

For the calculation of "control points" (coefficients) in the spline representations of T_1 and F_1, we herein employ a collocation scheme using as collocation points ($\{\tau_j\}, j = 0, \ldots, n+m$) the Greville abscissae associated with the refined knotvector $\mathscr{T}^{(m)}$. Hence, by collocating at τ_j, we get the following system of linear equations which we need to solve with respect to the unknown coefficients $T_{1,i}$ and $F_{1,i}, i = 0, \ldots, n+m$:

$$\sum_{i=0}^{n+m} F_{1,i} \left(\int_0^1 G_1(t,\tau_j) R_{i,k}^{(m)}(t) \|\dot{\mathbf{p}}(t)\| dt \right) -$$

$$-\sum_{i=0}^{n+m} T_{1,i} \left(\int_0^1 R_{i,k}^{(m)}(t) \frac{\partial G_1(t,\tau_j)}{\partial \mathbf{n}_1(t)} \|\dot{\mathbf{p}}(t)\| dt + \frac{R_{i,k}^{(m)}(t)}{2} \right) = 0,$$

$$j = 0, \ldots, m+n \quad (13a)$$

$$-\frac{\kappa_1}{\kappa_2} \sum_{i=0}^{n+m} F_{1,i} \left(\int_0^1 G_2(t,\tau_j) R_{i,k}^{(m)}(t) \|\dot{\mathbf{p}}(t)\| dt \right) +$$

$$+\sum_{i=0}^{n+m} T_{1,i} \left(\int_0^1 R_{i,k}^{(m)}(t) \frac{\partial G_2(t,\tau_j)}{\partial \mathbf{n}_1(t)} \|\dot{\mathbf{p}}(t)\| dt - \frac{R_{i,k}^{(m)}(t)}{2} \right) =$$

$$= \frac{1}{2} - \int_0^1 \frac{\partial G_2(t,\tau_j)}{\partial \mathbf{n}_1(t)} \|\dot{\mathbf{p}}(t)\| dt, \quad j = 0, \ldots, m+n \quad (13b)$$

Obviously, the solution of the linear system in Eqs. 13a,13b comprises the "control points" of the temperature $\{T_{1,i}\}$ and heat flux $\{F_{1,i}\}$ fields, $i = 0, \ldots n+m$ along the separating interface, respectively. Plugging this heat flux representation in Eq. 6 results in the following relation used for the calculation of heat transfer along S for one period L:

$$h_T = \int_S \frac{\partial T_1(\mathbf{p})}{\partial \mathbf{n}_1(\mathbf{p})} d\ell(\mathbf{p}) = \sum_{i=0}^{n+m} F_{1,i} \int_0^1 R_{i,k}^{(m)}(t) \|\dot{\mathbf{p}}(t)\| dt \quad (14)$$

The Green functions appearing in both first terms of Eqs. 13a,13b exhibit logarithmic singularities when $t = \tau_j$ as can be easily observed in Eqs. 8,9. However, these singularities are eliminated when t does not belong to the local support of $R_{i,k}^m(t)$. Hence, logarithmic singularities are only exhibited when t lies in the local support of the corresponding basis function. Furthermore, as it can be easily deduced, the gradient of Green's functions also exhibits a singularity; however the integrands of the second terms in Eqs. 13a,13b have no singularities as the normal derivative is regular under the assumption of S being a sufficiently smooth curve; see [3] and [1]. We calculate all integrals by firstly eliminating parts where the basis functions is zero and splitting the remaining parts so that singularities, when they occur, fall on integration boundaries. Subsequently, applying an adaptive Gauss-Kronrod quadrature scheme, allows us to successfully compute their values, as this scheme guarantees integration of up to square root singularities at boundaries; see [34].

2.3 IGABEM Solver Efficiency

The IGABEM approach described in the previous section (Sect. 2.2) has been extensively tested in [17] with respect to its efficiency and accuracy in estimating the values of heat transfer, heat flux and temperature distributions for the problem presented in Sect. 2. The authors selected several different separating interface geometries and reference solutions were acquired with the use of the commercial finite element computational package COMSOL [7] employing extremely fine meshes of around 6×10^4 elements for each case. It was demonstrated that 48 degrees of freedom, i.e., the number of control points used in interface's representation, suffice in achieving an error of less than 10^{-6} for both heat transfer values and heat flux distributions in all examined cases. Further to this, convergence rates have been computed and compared against the convergence rates achieved by a standard BEM approach described in [9]. Specifically, for the case of the IGABEM approach, convergence rates were demonstrated to reach $O(n^{-4})$ in all cases against linear convergence rates for the low-order standard BEM. In practice, the low standard BEM approach requires around 10^3 DoFs against only 48 for IGABEM to achieve the same level of accuracy in solution's estimation.

Finally, apart from the obvious advantage in the number of required DoFs, Kostas et al. in [17] demonstrated that the IGABEM approach is also advantageous with regards to computing time. Specifically, Kostas et al. presented a plot (Fig. 7 in [17]) of required computing time against achieved accuracy for the two BEM approaches. Using that graph, one can easily see that a maximum deviation of 10^{-6} from the reference solution requires around 150 seconds for the IGABEM approach while 10^4 secs would be needed for the low-order standard BEM [9].

3 Parametric Model

A series of parametric models have been developed for the generation of separating interface's shape instances. These range from simplified versions that directly define control points' coordinates to more elaborate ones that employ a number of affine transformations for the definition of interface's geometrical shape. In all cases, the generation process guarantees a tangent continuous, non self-intersecting periodic separating interface shape that splits the domain in question in exactly two separate domains. Depending on the application under consideration, additional conditions such as linear relationships between domains' areas and/or interface's length constraints may be applied. One parameter that is present in all parametric models is the number of control points ($N, N \geq 4$) used for representing the interface shape. Hence, in all cases, the parametric model comprises an appropriate process for defining the required $2 \times N$ control point coordinates[1] from a set of parameters with a cardinality, which ideally is significantly less than $2 \times N$. For reasons of simplicity and without loss of generality, we may assume here that the basic dimensions of our bilayered structure period are $H = L = 1$.

Finally, in all parametric models discussed in the sequel, we have to satisfy several side constraints imposed on the design parameters, e.g., $v_i \in [a_i, a_{i+1}]$ and $v_{i+1} \in [a_{i+1}, a_{i+2}]$ with $a_i < a_{i+1} < a_{i+2}$. An easy and robust way for satisfying these side constraints is to employ an affine transformation from $[0, 1]$ to $[a_i, a_{i+1}]$ and therefore define a different set of parameters $\{\bar{v}_i\}$ that are strictly in $[0, 1]$ and define $\{v_i\}$ as follows:

$$v_i = \bar{v}_i(a_{i+1} - a_i) + a_i, \quad \text{if } v_i \in [a_i, a_{i+1}].$$

Obviously, if we want to make v_i lie in the corresponding open interval, i.e., $v_i \in (a_i, a_{i+1})$, we may still define $\bar{v}_i$ to lie in $[0, 1]$ and define the transformation as follows:

$$v_i = \bar{v}_i(a_{i+1} - a_i - 2\epsilon) + a_i + \epsilon, \quad \text{if } v_i \in (a_i, a_{i+1}), \ 1 >> \epsilon > 0.$$

3.1 Uniformly Distributed Control Points

This is the simplest of the developed parametric models and interface shape generation is accomplished via a design vector of $N - 2$ components. Specifically, we assume that the interface is represented as a cubic B-Spline curve with N control points and a clamped knotvector. The first and last control points should obviously need to be:

$$\mathbf{b}_0 = (b_{0,x}, b_{0,y}) = (0, v_0), \quad \mathbf{b}_{N-1} = (b_{N-1,x}, b_{N-1,y}) = (1, v_0). \tag{15}$$

where v_0 is our first parameter with $0 + \epsilon \leq v_0 \leq 1 - \epsilon$, $1 >> \epsilon > 0$. We further assume that the remaining control points, with the exception of the second and next

---

[1] As we limit our investigation in 2D curves, each control point can be represented by a 2D point.

to last one, are uniformly distributed in (0, 1) in the horizontal direction. In other words

$$b_{i+1,x} = \frac{i}{N-3}, \quad i = 1, \ldots N - 4.$$

For the second point, $\mathbf{b}_1$ we may assume that $b_{1,x} = \frac{1}{2(N-3)}$, or, more generally, $b_{1,x} \in (0, \frac{1}{N-3})$, while the next to last control point will be defined through the tangential direction[2] equality equation:

$$\frac{b_{1,y} - b_{0,y}}{b_{1,x} - b_{0,x}} = \frac{b_{y,N-1} - b_{y,N-2}}{b_{x,N-1} - b_{x,N-2}}. \tag{16}$$

Therefore, the remaining $N - 3$ parameters $(v_1, \ldots, v_{N-3})$ with $v_i \in (\epsilon, 1 - \epsilon)$, correspond to the y-coordinates of the control points $\{\mathbf{b}_1, \ldots, \mathbf{b}_{N-3}\}$.

Let us examine a simple case with 8 control points ($N = 8$). The required parameters values are obviously 6. If we further assume that $\{v_i\} = \{0.2, 0.1668, 0.9606, 0.941, 0.4902, 0.794\}$ we will get a separating interface instance represented by a cubic B-Spline curve $\mathbf{c}(u) = \sum_{i=0}^{7} \mathbf{b}_i N_{i,4}(u)$ (see Fig. 2) with

$$\mathbf{b}_0 = (0, v_0), \mathbf{b}_1 = \left(\frac{1}{10}, v_1\right), \mathbf{b}_2 = \left(\frac{1}{5}, v_2\right), \mathbf{b}_3 = \left(\frac{2}{5}, v_3\right), \mathbf{b}_4 = \left(\frac{3}{5}, v_4\right),$$

$$\mathbf{b}_5 = \left(\frac{4}{5}, v_5\right), \mathbf{b}_6 = \left(\frac{9}{10}, v_0 - (v_1 - v_0)\right), \mathbf{b}_7 = (1, v_0) \Rightarrow$$

$$\mathbf{b}_0 = (0, 0.2), \mathbf{b}_1 = (0.1, 0.1668), \mathbf{b}_2 = (0.2, 0.9606), \mathbf{b}_3 = (0.4, 0.941),$$

$$\mathbf{b}_4 = (0.6, 0.4902), \mathbf{b}_5 = (0.8, 0.794), \mathbf{b}_6 = (0.9, 0.2332), \mathbf{b}_7 = (1, 0.2).$$

For the calculation of $\mathbf{b}_{N-2}$ we employ the following approach:

- We first assume that $b_{x,N-2} = \frac{2N-7}{2(N-3)}$ and then calculate $b_{y,N-2}$ using Eq. 16.
- If $b_{y,N-2} < \epsilon$, we set $b_{y,N-2} = 0$ and solve Eq. 16 with respect to $b_{x,N-2}$.
- Similarly, if $b_{y,N-2} > 1 - \epsilon$, we set $b_{y,N-2} = 1$ and solve Eq. 16 with respect to $b_{x,N-2}$.

It is quite clear that this simplistic approach imposes a restriction on the interface shapes we may generate since the longitudinal positions of control points are fixed and their values follow a strict ascending order. The former restriction can be easily eliminated by allowing longitudinal coordinates to vary while satisfying the

---

[2] We do not impose equality of magnitudes and hence we only require 1st-order geometric continuity, i.e., G^1.

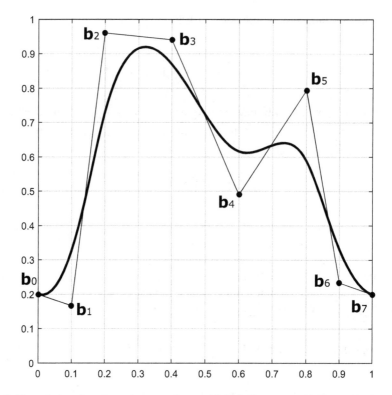

Fig. 2 Example interface shape represented as a cubic B-Spline curve with 8 control points

following rule:

$$b_{i+1,x} \in \left[\frac{2i-1}{2(N-3)}, \frac{2i+1}{2(N-3)} \right], \quad i = 1, \ldots, N-4 \tag{17}$$

If we extend our model as mentioned above, we will need to introduce additional parameters in the design vector that will determine the longitudinal position of these $N - 4$ control points. Therefore, instead of moving on a vertical line placed at $b_{i,x}$, control point b_i will assume a position in the rectangular domain $\left[\frac{2i-3}{2(N-3)}, \frac{2i-2}{2(N-3)} \right] \times [\epsilon, 1 - \epsilon]$. Although this is a very simple extension, the cardinality of the design parameters' set, in this case, gets close to $2N$ and consequently the design space increases significantly in dimension. This has an obvious negative effect on the complexity of the optimization problem we will need to solve and should be generally avoided.

For partially eliminating the latter restriction, i.e., allowing longitudinal coordinates to assume non-ascending order, but still produce a non self-intersecting separating interface shape, we may follow a procedure as described in the following section.

3.2 Protrusions and Interfaces with Fixed Parts

This parametric model may be employed if we want to model a protrusion of one medium into the other (or, equivalently, a recession of the latter medium with respect to the first one). Here we may additionally assume that a part of the separating interface is fixed and only the remaining part is allowed to change. As with the previous parametric model, two variants are available: one that limits the positioning of control points along a line segment and a more general one, which allows positioning within a convex polygon; see dashed lines and shaded areas in Fig. 3, respectively.

The first variant of this parametric model uses a small number of parameters, i.e., $N - 2$, and the two pairs of boundary control points, i.e., $\mathbf{b}_0, \mathbf{b}_1, \mathbf{b}_{N-2}$ and $\mathbf{b}_{N-1}$, are determined by the same parameters, v_0 and v_1, we have introduced in the previous section. Obviously, for the calculation of their coordinates, Eqs. 15,16 are employed once again. The remaining parameters, $v_2, \ldots, v_{N-3} \in [0, 1]$, are used to determine the position of each $\mathbf{b}_i = (1 - v_i)\mathbf{p}_i + v_i \mathbf{q}_i$ on the line segment (ray) defined by $\mathbf{p}_i, \mathbf{q}_i$; see Fig. 3. These points, $\mathbf{p}_i$ and $\mathbf{q}_i$, are defined as the midpoints of $\overline{\mathbf{p}_{a,i}\mathbf{p}_{b,i}}$ and $\overline{\mathbf{q}_{a,i}\mathbf{q}_{b,i}}$, respectively. Finally, $\overline{\mathbf{p}_{a,i}\mathbf{p}_{b,i}}$ and $\overline{\mathbf{q}_{a,i}\mathbf{q}_{b,i}}$, are defined by uniformly dividing the line defined by C_1, C_4 and the polygonal line defined by C_1, C_2, C_3, C_4 in $N - 4$ segments, respectively; see Fig. 3a.

This model can be extended to allow control points to assume position in shaded domains depicted in Fig. 3 with the introduction of additional parameters that will control the position of $\mathbf{p}_i$ and $\mathbf{q}_i$ within $\overline{\mathbf{p}_{a,i}\mathbf{p}_{b,i}}$ and $\overline{\mathbf{q}_{a,i}\mathbf{q}_{b,i}}$, respectively. Obviously, as in the case of the parametric model discussed in Sect. 3.1, this

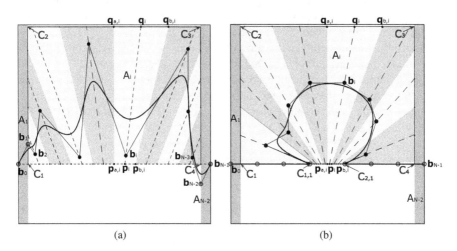

Fig. 3 Example interface shapes representing protrusions of one medium into the other. (**a**) Modeling a protrusion over the whole period. (**b**) Interface with a fixed part

extension will explode the number of required parameters and therefore increases the complexity of the corresponding optimization model.

Finally, we may use the same parametric model to restrict the part of the interface that is allowed to change. Specifically, using points $C_{1,1}$ and $C_{2,1}$ shown in Fig. 3b we can keep the parts $\overline{C_1, C_{1,1}}$ and $\overline{C_{2,1}, C_2}$ fixed, and restrict the emanation of rays in $\overline{C_{1,1}, C_{2,1}}$. This functionality can be used for the generation of protrusions as it is demonstrated in the heat maximization examples presented in Sect. 4.1.2.

4 Optimization Environment and Examples

The developed optimization framework comprises three components: the optimizer, the simulator and the parametric modeler. The framework is fully automated and the optimization procedure initiates by the optimizer component and terminates when an optimum solution has been found or when user-specified termination criteria have been met. These three components have been embedded within appropriate wrappers that guarantee the component interconnectivity and establish standard interface templates, which allow the replacement of the algorithm and/or tool employed in each component without any need for adjustments and/or modifications of the optimization framework and the way it is invoked.

All components have been implemented in MATLAB programming environment [21] with the aid of the NURBS [37], optimization [23] and global optimization [22] toolboxes. The optimizer component has been successfully tested with both global and local optimization algorithms and the examples presented in sequel utilize either the Genetic Algorithm implementation found in [22] or a local, gradient-based algorithm (Interior Point algorithm; see [34]) included in MATLAB's optimization toolbox function fmincon; see [23].

Generally, global optimization algorithms, such as Genetic Algorithms, Simulated Annealing, Particle Swarm Optimization and others, do not rely on a good starting point for reaching the global optimum. They are very well suited for estimating, with very high probability, the global optimum in highly non-linear objective functions (as the ones employed in our examples). Their main drawback is the requirement of a generally large number of iterations before reaching convergence. On the other hand, local, gradient- and/or Hessian-based algorithms require a good initial guess for reaching the global optimum and an appropriate starting point will allow the method to converge rapidly to the global optimum. However, when we cannot provide an appropriate starting point, e.g., the optimum shape cannot be estimated, the method will probably reach a local minimum/maximum and hence, the use of guided random search algorithms is generally preferred. Details for the usage of local and global optimization schemes are included in the discussion of the examples presented later in this section.

The parametric modeler component utilizes the models presented in Sects. 3.1 and 3.2 along with their variants described in the same subsections. The parametric modeler generates a cubic NURBS representation of the separating interface for a given design vector **v**. The specific details of the required parameters and model setup are also provided in the discussion of the examples that follow.

Finally, the simulator component comprises the implementation of the IGABEM approach described in Sect. 2. The simulator interface allows the user to specify the required temperatures and dimensions, conductivity coefficients along with the number of additional knots (m) that may be inserted; see the relevant discussion before Eqs. 10, 11. Knot insertions refine the spline space used for the representation of the unknown quantities (temperature and heat flux) and therefore, more accurate solutions can be achieved.

In the remaining part of this section we will present indicative examples for both forward and inverse design cases that demonstrate the way to acquire answers for the questions posed in Sect. 1.1. In most cases, we will obviously need to introduce additional design constraints to avoid getting trivial solutions or having a not well defined problem. We will begin our presentation with forward design cases where the complete length or part of the separating interface is determined by a shape optimization procedure that maximizes the heat transfer between the two conductive media and continue with inverse design and the determination of the interface shape that guarantees a given heat transfer value or a specific temperature distribution along its length.

4.1 Forward Design

When examining the forward design family of problems, it is quite clear that we need to set some additional design constraints since otherwise, our heat transfer maximization problem will obviously attain a trivial solution that places the separating interface as close as possible to the high temperature domain boundary. The design constraints we have employed are simple geometric constraints that correspond to practical design restrictions and application requirements.

Specifically, we will demonstrate the results acquired via the shape optimization framework for three different cases:

1. the separating interface shape that maximizes heat transfer when the volume of the two media needs to be fixed. Based on our 2D problem setup, this requirement corresponds to a constraint on the area values of the domains Ω_1, Ω_2; see Fig. 1.
2. The separating interface shape that maximizes heat transfer under a given surface area value for the separating interface. In this case, the design constraint, for our 2D case setup, corresponds to a restriction on the length of the separating interface.
3. Finally, we demonstrate solutions for a heat-transfer-maximizing interface when both media volume and interface surface area are constrained.

4.1.1 Heat Transfer Maximization Subject to Area Constraints

For the first case we will present two different examples demonstrating the use of our framework for determining the heat-transfer-maximizing interface under given media volumes. For the first case, we assume that we are given an initial straight-line separating interface that splits our problem domain (Ω) into two pieces Ω_1 and Ω_2 with equal area values; see Fig. 4. Obviously, within within the period $L = 1$, each domain has an area of 0.5 square units and our optimization goal is to modify the shape of the separating interface so that we achieve heat transfer maximization without affecting the domain area equality. Furthermore, we assume that both the top and bottom surface of the bilayered construction are isothermal boundaries with temperatures 1 K ($\tilde{T}_2 = T_2 - c_2 = 1 - 1$) and 0 K ($\tilde{T}_1 = T_1 - c_1 = 0 - 0$), respectively. Finally, the conductivity coefficients for the top (medium 2) and bottom medium (medium 1) are $\kappa_2 = 0.1 \frac{W}{mK}$ and $\kappa_1 = 1 \frac{W}{mK}$, respectively.

The parametric model employed in this case is the simple model with uniformly distributed control points described in Sect. 3.1. Specifically, we have assumed a cubic B-Spline curve with a uniform, clamped knotvector and 12 control points (24 coordinate values) for the representation of the separating interface shape. Therefore, the required number of parameters according to our approach will be equal to $N - 2 = 10$. Further to this, we assume here a horizontal tangent direction at both ends of the periodic interface curve, which simplifies the problem and

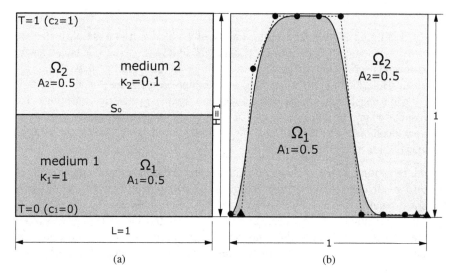

Fig. 4 Example 1: Interface shape optimization under area constraints. (**a**) Initial separating interface. (**b**) Optimum separating interface; circles and triangles denote free and dependent control points, respectively

decreases the number of required parameters as shown below:

$$\mathbf{b}_0 = (0, v_0) \quad \mathbf{b}_1 = (\tfrac{1}{18}, v_0) \quad \mathbf{b}_2 = (\tfrac{1}{9}, v_1) \quad \mathbf{b}_3 = (\tfrac{2}{9}, v_2)$$
$$\mathbf{b}_4 = (\tfrac{3}{9}, v_3) \quad \mathbf{b}_5 = (\tfrac{4}{9}, v_4) \quad \mathbf{b}_6 = (\tfrac{5}{9}, v_5) \quad \mathbf{b}_7 = (\tfrac{6}{9}, v_6)$$
$$\mathbf{b}_8 = (\tfrac{7}{9}, v_7) \quad \mathbf{b}_9 = (\tfrac{8}{9}, v_8) \quad \mathbf{b}_{10} = (\tfrac{17}{18}, v_0) \quad \mathbf{b}_{11} = (1, v_0),$$

where $v_i \in (0, 1)$, $i = 0, \ldots, 8$. Hence, the initial straight line interface obviously corresponds to the design vector $\mathbf{v} = \{v_i\}$, $v_i = 0.5$, $\forall i$. We need to note here that the number of control points obviously affects the parametric model's ability to capture the optimum shape. Specifically, if the optimum interface has a complex shape, we will not be able to generate its shape with a small number of control points. On the other hand, a large number of control points (and therefore parameter values) increases the dimensions of the design space and renders the optimization procedure computationally expensive. A general rule of thumb for the cases examined is to start with around 10 control points and increase them until no further enhancement of the optimum shape can be achieved.

Under these assumptions, we can state our heat transfer maximization problem as follows:

$$\max h_T(\mathbf{v}) = \max \sum_{i=0}^{n+m} F_{1,i}(\mathbf{v}) \int_0^1 R_{i,k}^{(m)}(t) \|\dot{\mathbf{p}}(t; \mathbf{v})\| dt$$

subject to : \hfill (18)

$$A_1 = \int_0^1 p_y(t; \mathbf{v}) \dot{p}_x(t; \mathbf{v}) dt = 0.5,$$

$$v_i \in (0, 1), \; i = 0, \ldots, 8,$$

where the former constraint enforces the area equality for the two media and the latter guarantees that the interface will touch neither of the isothermal boundaries. In Fig. 4b we depict the acquired optimum interface. This optimum shape has been acquired with the help of MATLAB's Genetic Algorithm implementation. The same result may be reached with the aid of gradient- and/or Hessian-based optimization algorithms, assuming a good initial shape guess is provided. The resulting optimum shape achieves a heat transfer value $h_T = 0.469$, which is considerably larger than the value for the initial straight-line interface, $h_{T_0} \approx 0.1818$. We must note here that due to the nature of our problem (consideration of a periodic interface in an infinite bilayered structure) the result of the optimization procedure can be any horizontal translational transformation (and/or any other affine transformation, such as horizontal mirroring the will obviously produce the same infinite separating interface) of the shape depicted in Fig. 4b.

In the second optimization example, we once again begin with a straight line separating interface but this time, as depicted in Fig. 5a, the two domains have unequal thicknesses and consequently unequal areas. Assuming once again a height

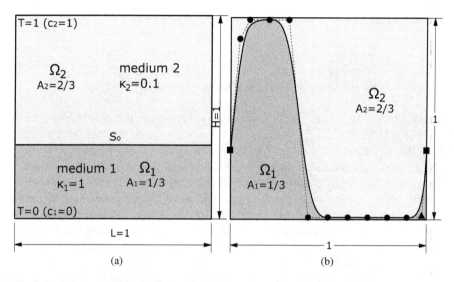

Fig. 5 Example 2: Interface shape optimization under area constraints. (**a**) Initial separating interface. (**b**) Optimum separating interface; circles denote free control points, squares and triangles denote fixed and dependent control points, respectively

of $H = 1$ for the bilayered structure, the two media will have an area of $\frac{1}{3}$ and $\frac{2}{3}$ square units within a period of $L = 1$, respectively. We employ the same parametric model using a cubic B-Spline curve with uniformly distributed control points for the representation of the separating interfaces. This time, we assume that the interface's endpoints are fixed at a height of $1/3$ but tangent directions there may deviate from the horizontal direction. Experimentation with different numbers of interface's control points has indicated that 13 control points suffice for reaching the optimum and therefore 10^3 parameter values $(v_0, \ldots, v_9)$ are required:

$$\begin{aligned}
&\mathbf{b}_0 = (0, \tfrac{1}{3}) \quad \mathbf{b}_1 = (\tfrac{1}{20}, v_0) \quad \mathbf{b}_2 = (\tfrac{1}{10}, v_1) \quad \mathbf{b}_3 = (\tfrac{2}{10}, v_2) \\
&\mathbf{b}_4 = (\tfrac{3}{10}, v_3) \quad \mathbf{b}_5 = (\tfrac{4}{10}, v_4) \quad \mathbf{b}_6 = (\tfrac{5}{10}, v_5) \quad \mathbf{b}_7 = (\tfrac{6}{10}, v_6) \\
&\mathbf{b}_8 = (\tfrac{7}{10}, v_7) \quad \mathbf{b}_9 = (\tfrac{8}{10}, v_8) \quad \mathbf{b}_{10} = (\tfrac{9}{10}, v_9) \quad \mathbf{b}_{11} = (b_{11x}, b_{12y}) \quad \mathbf{b}_{12} = (1, \tfrac{1}{3}),
\end{aligned}$$

where the coordinates of $\mathbf{b}_{11}$ need to satisfy the tangential continuity as defined by Eq. 16. The remaining boundary conditions and conductivity coefficients are identical to the ones used in the previous optimization example. Therefore, we can

---

[3] One less than $N - 2$ since endpoint heights are fixed.

state our heat transfer maximization problem as follows:

$$\max h_T(\mathbf{v}) = \max \sum_{i=0}^{n+m} F_{1,i}(\mathbf{v}) \int_0^1 R_{i,k}^{(m)}(t) \|\dot{\mathbf{p}}(t;\mathbf{v})\| dt$$

subject to : (19)

$$A_1 = \int_0^1 p_y(t;\mathbf{v}) \dot{p}_x(t;\mathbf{v}) dt = \frac{1}{3},$$

$$v_i \in (0,1), \ i = 0, \ldots, 9.$$

Similarly to the first optimization example, deterministic[4] and evolutionary algorithms may be employed in the solving process. In Fig. 5b, the result achieved using MATLAB's Genetic Algorithm implementation is depicted. The optimum heat transfer value is $h_T = 0.3397$, which is considerable larger when compared to the value acquired from the straight line interface $h_{T_0} \approx 0.142857$.

4.1.2 Heat Transfer Maximization Subject to Perimeter Constraints

In this second case of optimization examples, we present a series of results that demonstrate the use of the optimization framework in conjunction with the second parametric model, described in Sect. 3.2, for handling interfaces with protrusions. Specifically, we present a series of optimization examples for which a part of the periodic separating interface is assumed fixed while the remaining part is allowed to change. We could potentially use the same parametric model we have employed in the previous section but as it has been already shown that the optimum shapes will not be x-monotonic [9], the first parametric model will not be able to reach the optimum. We assume here that we have the same initial separating interface, i.e., a straight line interface splitting the bilayered structure periodic domain in two domains Ω_1 and Ω_2 as depicted in Fig. 5a. The remaining boundary conditions, conductivity coefficients and temperature values are identical to the 2nd example of the previous section and, in this case, we seek to generate a centrally-located protrusion that maximizes heat transfer.

In Fig. 6, we include the results attained when solving the optimum protrusion problem subject to two major constraints: (a) a restriction on the length (perimeter P) of the interface and (b) a restriction on the percentage of the initial interface that is allowed to change. Specifically, we vary the allowable percentage from 20% to 90% (black to green family of curves in Fig. 6) and the maximum allowable perimeter from 1.2 to 2 units of length (dotted to solid line curves in the same figure).

---

[4] Assuming a good initial estimation of the optimum shape can be made.

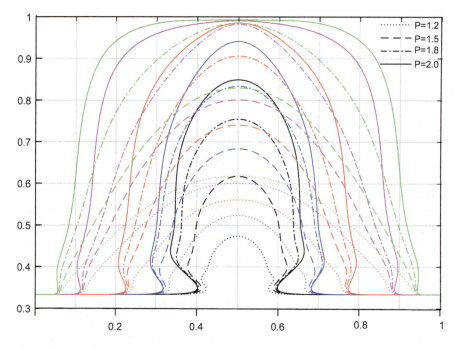

Fig. 6 Example 3: Optimum protrusions

The family of heat transfer maximizing interfaces shown in Fig. 6 are obtained by solving the following series of problems:

$$\max h_T(\mathbf{v}) = \max \sum_{i=0}^{n+m} F_{1,i}(\mathbf{v}) \int_0^1 R_{i,k}^{(m)}(t) \|\dot{\mathbf{p}}(t;\mathbf{v})\| dt$$

subject to : (20)

$$\int_0^1 \|\dot{\mathbf{p}}(t;\mathbf{v})\| dt \leq P, \quad P \in \{1.2, 1.5, 1.8, 2.0\}$$

$$v_i \in (0,1), \quad \forall i.$$

Let us note here that, in this case, we can easily employ MATLAB's deterministic algorithms, such as the Interior Point algorithm [5] implemented in MATLAB's fmincon function, for seeking the optimum interface, since the estimation of a good initial shape is rather obvious in this case.

4.1.3 Heat Transfer Maximization Subject to Area and Perimeter Constraints

The last set of examples examined for the forward design case comprises a set of problems employing both area and interface-perimeter constraints. The interface parametric model used in this case is the same simple parametric model with the same parameters and assumptions employed for the 2nd example in Sect. 4.1.1 and the initial separating interface is as depicted in Fig. 5a.

The family of heat transfer maximizing interfaces shown in Fig. 7 is obtained by solving the following series of problems:

$$\max h_T(\mathbf{v}) = \max \sum_{i=0}^{n+m} F_{1,i}(\mathbf{v}) \int_0^1 R_{i,k}^{(m)}(t) \|\dot{\mathbf{p}}(t;\mathbf{v})\| dt$$

subject to : (21)

$$\int_0^1 \|\dot{\mathbf{p}}(t;\mathbf{v})\| dt \leq P, \quad P \in \{1.2, 1.5, 1.8, 2.0\}$$

$$A_1 = \int_0^1 p_y(t;\mathbf{v}) \dot{p}_x(t;\mathbf{v}) dt = \frac{1}{3},$$

$$v_i \in (0, 1), \ \forall i.$$

Similarly to the examples and the discussion of optimum fins presented in [18] and [9] a wavy separating interface is formed in all cases, where longer and shallower valleys, and deeper and narrower protrusions of the high conductivity material into the lower conductivity material are developed as the perimeter inequality constraint tends to infinity. Obviously, when $P = \infty$, the problem in Eq. 21 becomes essentially the same problem described in Eq. 19 and the interface consequently assumes the shape shown in Fig. 5b. Finally, the obtained heat transfer values increase following the allowable perimeter length, i.e., $h_T = \{0.1429, 0.1533, 0.1727, 0.1957, 0.2181, 0.3397\}$ when $P \leq \{1, 1.2, 1.5, 1.8, 2, \infty\}$, respectively; see Figs. 5 and 7.

4.2 Inverse Design

In this second family of problems, we employ the same shape optimization framework for addressing inverse design problems. Specifically, the examples presented in this section refer to design solution for the following cases:

1. What is the shape of the periodic separating interface that achieves a given heat transfer value between the two media in our bilayered structure?

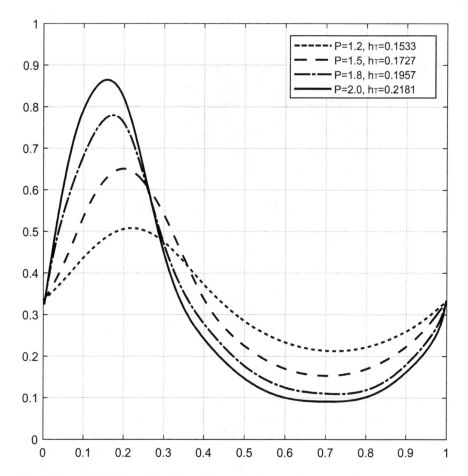

Fig. 7 Example 4: Optimum interfaces subject to perimeter and area constraints

2. What is the shape of the periodic separating interface that corresponds to a given temperature (or heat flux) distribution along an unknown interface of our bilayered structure?

As we will discuss in more detail in the subsequent examples, the first question will not assume a unique solution in the general case while the second question will generally do, assuming obviously that a feasible solution exists.

4.2.1 Interface Shape for a Given Heat Transfer Value

In this first series of inverse design example cases, we assume a given heat transfer value while the shape of the separating interface constitutes the unknown quantity. To facilitate the comparison with the examples presented in the previous section, we assume, once again, the same boundary value problem and setup as described for the 2nd example presented in Sect. 4.1.1, including the constraint of fixed interface endpoints at a height equal to 1/3. The parametric model employed for all cases in this subsection is the simple model with uniformly distributed control points described in Sect. 3.1. This time we assume 14 control points for the interface's cubic B-Spline representation, which are defined with the aid of 11 parameters ($\{v_i\}$, $i = 0,$):

$$\mathbf{b}_0 = (0, \tfrac{1}{3}) \quad \mathbf{b}_1 = (\tfrac{1}{22}, v_0) \quad \mathbf{b}_2 = (\tfrac{1}{11}, v_1) \quad \mathbf{b}_3 = (\tfrac{2}{11}, v_2) \quad \mathbf{b}_4 = (\tfrac{3}{11}, v_3)$$
$$\mathbf{b}_5 = (\tfrac{4}{11}, v_4) \quad \mathbf{b}_6 = (\tfrac{5}{11}, v_5) \quad \mathbf{b}_7 = (\tfrac{6}{11}, v_6) \quad \mathbf{b}_8 = (\tfrac{7}{11}, v_7)$$
$$\mathbf{b}_9 = (\tfrac{8}{11}, v_8) \quad \mathbf{b}_{10} = (\tfrac{9}{11}, v_9) \quad \mathbf{b}_{11} = (\tfrac{10}{11}, v_{10}) \quad \mathbf{b}_{12} = (b_{12x}, b_{12y}) \quad \mathbf{b}_{13} = (1, \tfrac{1}{3}),$$

where the coordinates of $\mathbf{b}_{12}$ need to satisfy the tangential continuity as defined by Eq. 16. The optimization problem that we will need to solve in all cases is the following:

$$\min \left| \sum_{i=0}^{n+m} F_{1,i}(\mathbf{v}) \int_0^1 R_{i,k}^{(m)}(t) \|\dot{\mathbf{p}}(t; \mathbf{v})\| dt - h_{T_j} \right|, \ j = 1, 2, 3 \quad (22)$$

subject to :

$$v_i \in (0, 1), \ i = 0, \ldots, 10,$$

where $h_{T_1} = 0.2, h_{T_2} = 0.3397$ and $h_{T_3} = 0.5$. Before we proceed with the presentation of our results, let us note here that the first given heat transfer value is less than maximum attainable[5] heat transfer value, the second one coincides with it, while the last once (h_{T_3}) is larger; see Fig. 5b and the corresponding discussion in Sect. 4.1.1.

Figure 8 depicts indicative results for the separating interface shape acquired by the solution of Eq. 22 for $h_{T_1} = 0.2$. Two different initial estimations have been used (see dashed lines in Figs. 8a,b) in conjunction with MATLAB's Interior Point algorithm implementation. In both cases, we achieve a solution that minimizes the objective function in Eq. 22 but obviously does not correspond to the same separating interface shape. As a matter of fact, we can have an infinite number of shapes that satisfy our objective function and therefore this problem does not admit a unique solution. This is not an issue related to the optimization algorithm employed as guided random search optimization algorithms will generally produce a different solution with each run and therefore this is an issue inherent to the problem under consideration. Additional constraints and/or problem setup modifications are required, if we want to achieve uniqueness in the solution.

However, if the given heat transfer value is greater or equal to the maximum attainable value for the specific bilayered structure setup, the solution becomes

---

[5]For the specific bilayered structure layout and problem setup.

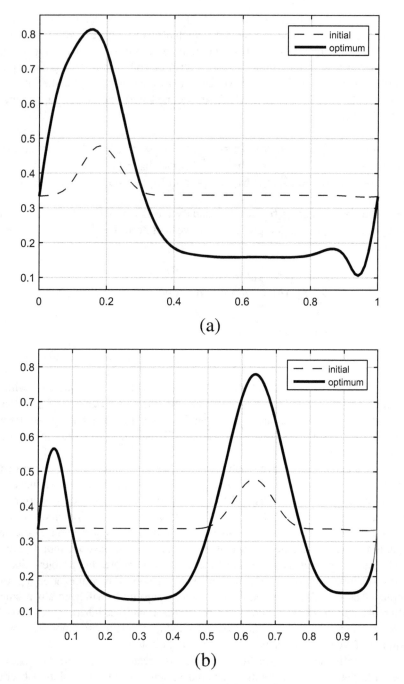

Fig. 8 Interface shape for a given heat transfer value of $h_{T_1} = 0.2$ using different initial shape estimates. (**a**) Objective function value 2.7×10^{-9}. (**b**) Objective function value 6×10^{-12}

unique[6] as it is depicted in Figs. 9 and 10. Of course this does not mean that we can obtain this unique optimum solution without a good initial estimate of interface's shape when gradient- and/or Hessian-based optimization algorithms are employed, as it is clearly depicted in Figs. 9a and 10a. The optimum and unique solution of Eq. 22 can be achieved with either a good initial shape estimation (see Figs. 9b and 10b) or by employing a global optimization approach with algorithms such as simulated annealing, evolutionary algorithms etc.

We can easily see in Fig. 9b that the result (thick solid line) of the inverse problem for $h_{T_2} = 0.3397$ practically coincides with the result of the corresponding forward problem's solution (included as a thin solid line in Fig. 9b). Although one may initially think that there is a difference in the narrow protrusion part depicted in the same figure, it is easy to observe that this deviation corresponds merely to a horizontal mirroring of the protrusion with respect to a vertical axis passing through the intersection point of the two curves and therefore has practically no effect on the heat transfer value.

Finally, Fig. 10b depicts the solution for $h_{T_3} = 0.5$ which is obviously identical to the optimum solution obtained for h_{T_2} and depicted in Fig. 9b. Figure 10a demonstrates the dependence of the gradient-based optimization algorithms from a good initial guess in reaching the global optimum and we can generally state that with the exception of cases where the optimum shape can be easily estimated, we suggest the use of guided random search algorithms, such as Genetic Algorithms, Simulated Annealing, Particle Swarm Optimization etc., which do not rely on a good starting point for reaching the global optimum.

4.2.2 Interface Shape for a Given Temperature Distribution

The last two examples, for the inverse design case, demonstrate the use of the optimization framework in identifying the separating interface shape that exhibits a given temperature distribution along it. The exact same setup and formulation can be also used if, instead of a temperature distribution, the heat flux distribution is given along the unknown interface. We assume here that a set of temperature (or heat flux) values is given along the unknown separating interface and we want to employ our optimization framework for identifying the shape of the interface that produces the given distribution.

For the two examples presented in this section and for simplicity's sake, we have further assumed that the unknown interface exhibits x-monotonicity and therefore we may assume that the distribution can be represented by a set of points where temperature values are the ordinates and x-coordinates their abscissae. In both examples we employ the same simple parametric model described in Sect. 3.1 driven by a design vector with 10 parameter ($\{v_i\}$, $i = 0, \ldots, 10$). Hence, employing

---

[6]In the sense discussed in Sect. 4.1.1 where translational and/or mirroring transformations of the resulting shape are producing the same infinite periodic interface.

Fig. 9 Interface shape for a given heat transfer value of $h_{T_2} = 0.3397$ using different initial shape estimates. (**a**) Objective function value 0.0487; optimum solution missed. (**b**) Objective function value 1.8×10^{-9}; optimum solution reached

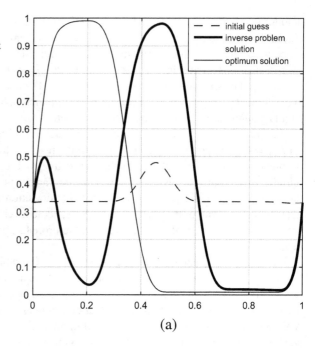

(a)

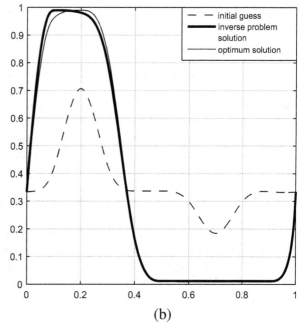

(b)

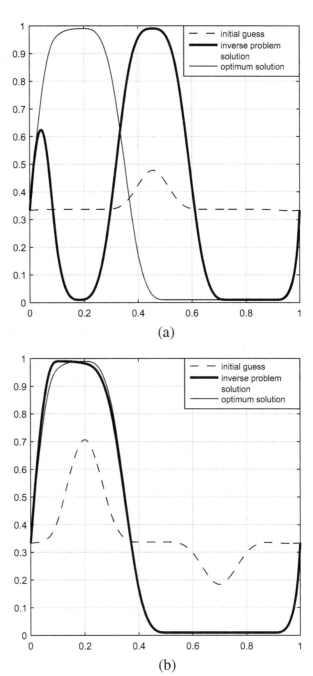

Fig. 10 Interface shape for a given heat transfer value of $h_{T_3} = 0.5$ using different initial shape estimates. (**a**) Objective function value 0.179; optimum solution missed. (**b**) Objective function value 0.1578; optimum solution reached

our IGABEM solver, we can identify the design vector that produces a similar temperature distribution and therefore identify the shape of the interface. For doing so, we need to minimize the distance between the computed and given temperature distribution and one way to do so is by minimizing the sum of the euclidean distances between corresponding 'temperature' points:

$$\min \left(\sum_{i=1}^{M} \left(T_1(t(x_i); \mathbf{v}) - T_0(x_i) \right)^2 \right)^{\frac{1}{2}}, \qquad (23)$$

$$v_i \in (0, 1), \ i = 0, \ldots, 9.$$

where M is a large number of points with a given temperature value $\{T_0(x_i)\}$, $i = 1, \ldots, M$ along the unknown separating interface.

Figure 11a depicts the given temperature distribution (dotted-line curve) for the optimum separating interface presented in Fig. 5b. The solid-line curve in the same figure corresponds to the solution of Eq. 23 using a simple gradient-based optimization algorithm. The actual shape of the interface is included in the same figure (depicted with thick black circular disks) and no visible difference exists between the original and the reconstructed separating interface shape; the minimum achieved objective function value for $M = 1000$ is equal to 1.05×10^{-6}.

Figure 11b depicts the given temperature distribution (dotted-line curve) for a randomly generated separating interface. The solid-line curve in the same figure corresponds to the solution of Eq. 23 using the same simple gradient-based optimization algorithm as in the previous example. The actual shape of the interface is included in the same figure (depicted once again with thick black circular disks) with a barely visible difference between the original and the reconstructed separating interface shape. This time the minimum achieved objective function value for $M = 1000$ is equal to 1.8×10^{-4}.

As we have already discussed in Sect. 4.1.1 the number of control points obviously affect the parametric model's ability to capture the separating interface's shape. If the temperature distribution exhibits a complex shape, we can safely assume that the interface shape will be equally complex and a sufficient number of parameters (control points) will be required for successfully reconstructing the unknown shape.

5 Conclusions and Future Work

In the present work, a robust design optimization framework has been developed and used for optimizing the shape of the periodic, separating interface between two infinite slabs in conduct with the aim of heat transfer maximization as described in Sects. 2.1 and 4.1. The same framework has been also employed for performing inverse design, i.e., finding the shape of the interface that generates a given heat

Fig. 11 Interface shape for a given temperature distribution. (**a**) Objective function value 1.05×10^{-6}. (**b**) Objective function value 1.8×10^{-4}

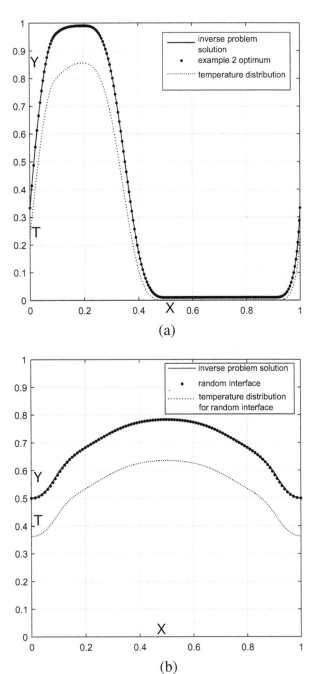

transfer and/or temperature distribution along itself; see Sect. 4.2. The design optimization environment is based on a series of geometric parametric modelers for generating instances of the separating interface, an IGABEM solver for calculating temperature and heat flux distributions along the interface, and an optimizer, which employs a series of local and global optimization algorithms.

The presented isogeometric boundary element method (IGABEM) is a central component in our design optimization framework and plays a significant role in permitting the automatic and efficient operation of the framework. This IGABEM solver is applied for solving the BIE system associated with the 2D steady-state heat transfer problem across a periodic interface separating two conducting and conforming media as described in Sect. 2.1. The isogeometric concept, in this context, is based on the exploitation of the same NURBS basis, used for the exact representation of interface's geometry, to approximate, via refinement, the physical quantities of temperature and normal heat flux along the interface. The enhanced accuracy and efficiency of this method has considerable benefits in its use for solving both forward and inverse design problems as has been demonstrated in Sect. 4.

Future work is planned towards further enrichment of the parametric models' capabilities and the extension of this framework for handling additional 2D and 3D heat transfer problem formulations. For the 3D case, one possibility includes the application of the doubly-periodic Green's function of the Laplace equation as described in Hautman and Klein [12], and Pozrikidis [29].

Acknowledgments This work has received funding from Nazarbayev University Grant No. 090118FD5328: "Shape Optimization of Lift and Thrust generating surfaces with the aid of IsoGeometric Analysis (SOLTIGA)"

References

1. Aimi A., Calabr F., Diligenti M., Sampoli M., Sangalli G., Sestini A. (2018) Efficient assembly based on B-spline tailored quadrature rules for the IGA-SGBEM, Comput. Methods Appl. Mech. Engrg. 331:327–342
2. An Z., Yu T., Bui T.Q., Wang C., Trinh N.A. (2018) Implementation of isogeometric boundary element method for 2-d steady heat transfer analysis, Adv. Eng. Softw. 116:36–49
3. Atkinson K.E. (1997) The Numerical Solution of Integral Equations of the Second Kind, Cambridge University Press
4. Belibassakis K.A., Gerosthathis T.P., Kostas K.V., Politis C.G., Kaklis P.D., Ginnis A.-A., Feurer C., A BEM-Isogeometric method for the ship wave-resistance problem, Ocean Eng. 60:53–67
5. Byrd R.H., Hribar M.E., Nocedal J. (1999) An Interior Point Algorithm for Large-Scale Nonlinear Programming, SIAM Journal on Optimization 9(4):877–900
6. Cho S., Ha S., Isogeometric shape design optimization:exact geometry and enhanced sensitivity, Structural and Multidisciplinary Optimization 38(1):53–70
7. COMSOL v.5.3 (2017) COMSOL Multiphysics User's Guide, https://www.comsol.com/
8. Cottrell J.A., Hughes T.J.R., Bazilevs Y. (2009) Isogeometric Analysis: Toward Integration of CAD and FEA, Wiley

9. Fyrillas M.M., Leontiou T., Kostas K.V. (2017) Optimum interfaces that maximize the heat transfer rate between two conforming conductive media, Int. J. Therm. Sci. 121:381–389
10. Ginnis A.I., Duvigneau R., Politis C., Kostas K.V., Belibassakis K., Gerostathis Th., Kaklis P.D. (2013) A Multi-Objective Optimization Environment for Ship-Hull Design based on a BEM-Isogeometric Solver, in fifth Conference on Computational Methods in Marine Engineering (Marine 2013), Hamburg, Germany
11. Ginnis A.I., Kostas K.V., Politis C.G., Kaklis P.D., Belibassakis K.A., Gerostathis Th.P., Scott M.A., Hughes T.J.R., Isogeometric boundary-element analysis for the wave-resistance problem using T-splines, Comput. Methods Appl. Mech. Engrg. 279:425–439.
12. Hautman J., Klein M.L. (1992) An ewald summation method for planar surfaces and interfaces, Mol. Phys. 75(2):379–395
13. Hughes T., Cottrell J., Bazilevs Y. (2005) Isogeometric analysis: CAD, finite elements, NURBS, exact geometry and mesh refinement, Computer Methods in Applied Mechanics and Engineering 194:4135–4195
14. Kaklis P.D., Politis C.G., Belibassakis K.A., Ginnis A.I., Kostas K.V., Gerostathis Th.P. (2017) Encyclopedia of Computational Mechanics, Fluids–Part 2 (Boundary-Element Methods and Wave Loading on Ships), second ed., Wiley, vol.2:1–35
15. Kostas K.V., Ginnis A.I., Politis C.G., Kaklis P.D. (2015) Ship-hull shape optimization with a T-spline based BEM-isogeometric solver, Comput. Methods Appl. Mech. Engrg. 284:611–622, Isogeometric Analysis Special Issue
16. Kostas K.V., Ginnis A.I., Politis C.G., Kaklis P.D., Shape-optimization of 2d hydrofoils using an isogeometric BEM solver, J. Comput. Aided Des. 82:79–87
17. Kostas K.V., Fyrillas M.M., Politis C.G., Ginnis A.I., Kaklis P.D. (2018) Shape optimization of conductive-media interfaces using an IGA-BEM solver, Comput. Methods Appl. Mech. Engrg. 340:600–614
18. Leontiou T., Kotsonis M., Fyrillas M.M. (2013) Optimum isothermal surfaces that maximize heat transfer. Int J Heat Mass Transf 263:13–9
19. Li K., Qian X. (2011) Isogeometric Analysis and Shape Optimization via Boundary Integral, Com- puter Aided Design 43(11):1427–1437
20. Lian H., Simpson R., Bordas S., Sensitivity Analysis and Shape Optimisation through a T-spline Isogeometric Boundary Element Method, in: International Conference on Computational Mechanics (CM13), Durham, UK, 2013.
21. MATLAB (2016) version 9.1.0. (R2016b) The MathWorks Inc.
22. MATLAB Global Optimization Toolbox (2016) Version 3.4.1 (R2016b) The MathWorks Inc.
23. MATLAB Optimization Toolbox (2016) version 7.5 (R2016b) The MathWorks Inc.
24. Nagy A.P., Abdalla M.M., Grdal Z. 2009, Isogeometric sizing and shape optimization of beam structures, in 50th AIAA/ASME/ASCE/AHS/ASC Structures, Structural Dynamics, and Materials Conference, Palm Springs, California, USA
25. Nguyen D., Evgrafov A., Gravesen J. (2012) Isogeometric shape optimization for electromagnetic scattering problems, Progress In Electromagnetics Research 45:117–146
26. Nørtoft P., Gravesen J. (2013) Isogeometric shape optimization in fluid mechanics, Structural and Multidisciplinary Optimization 48:909–925
27. Peake M., Trevelyan J., Coates G. (2013) Extended isogeometric boundary element method (XIBEM) for two-dimensional Helmholtz problems, Comput. Methods Appl. Mech. Engrg. 259:93–102
28. Politis C., Ginnis A., Kaklis P., Belibassakis K., Feurer C. (2009) An isogeometric BEM for exterior potential-flow problems in the plane, in 2009 SIAM/ACM Joint Conference on Geometric and Physical Modeling, 349–354
29. Pozrikidis C. (2000) Conductive mass transport from a semi-infinite lattice of particles, Int. J. Heat Mass Transfer 43(4):493–504
30. Pozrikidis C. (2002) A practical guide to boundary element methods with the software library BEMLIB. CRC Press
31. Qian X. (2010) Full analytical sensitivities in NURBS based isogeometric shape optimization, Computer Methods in Applied Mechanics and Engineering 199(29–32):2059–2071

32. Qian X., Sigmund O. (2011) Isogeometric Shape Optimization of photonic crystals via Coons patches, Computer Methods in Applied Mechanics and Engineering 200:2237–2255
33. Scott M.A., Simpson R.N., Evans J.A., Lipton S., Bordas S.P.A., Hughes T.J.R., Sederberg T. (2013) Isogeometric boundary element analysis using unstructured T-splines, Comput. Methods Appl. Mech. Engrg. 254(0):197–221
34. Shampine L.F. (2008) Vectorized adaptive quadrature in MATLAB, Comput. Appl. Math. 211:131–140
35. Simpson R., Bordas S., Trevelyan J., Rabczuk T. (2012) A two-dimensional Isogeometric Boundary Element Method for elastostatic analysis, Comput. Methods Appl. Mech. Engrg. 209–212:87–100
36. Simpson R.N., Scott M.A., Taus M., Thomas D.C., Lian H. (2014) Acoustic isogeometric boundary element analysis, Comput. Methods Appl. Mech. Engrg. 269:265–290
37. Spink D.M. (2010) NURBS Toolbox version 1.0.0.0. https://www.mathworks.com/matlabcentral/fileexchange/26390-nurbs-toolbox-by-d-m-spink. Updated 15 Jan 2010
38. Sun S., Yu T., Nguyen T., Atroshchenko E., Bui T. (2018) Structural shape optimization by IGABEM and particle swarm optimization algorithm, Eng. Anal. Bound. Elem. 88:26–40
39. Wall W., Frenzel M., Cyron C. (2008) Isogeometric structural shape optimization, Computer Methods in Applied Mechanics and Engineering 197(33–40):2976–2988

Isogeometric Methods for Free Boundary Problems

M. Montardini, F. Remonato , and G. Sangalli

Abstract We present in detail three different quasi-Newton isogeometric algorithms for the treatment of free boundary problems. Two algorithms are based on standard Galerkin formulations, while the third is a fully-collocated scheme. With respect to standard approaches, isogeometric analysis enables the accurate description of curved geometries, and is thus particularly suitable for free boundary numerical simulation. We apply the algorithms and compare their performances to several benchmark tests, considering both Dirichlet and periodic boundary conditions. Our results constitute a starting point of an in-depth analysis of the Euler equations for incompressible fluids.

1 Introduction

This work focuses on the isogeometric analysis (IGA) of free boundary problems. IGA, first presented in [10], is a recent extension of the standard finite element method where the unknown solution of the partial differential equation is approximated by the same functions that are adopted in computer-aided design for the parametrization of the problem domain. These functions are typically splines and

---

M. Montardini
Department of Mathematics, University of Pavia, Pavia, Italy
e-mail: monica.montardini01@universitadipavia.it

F. Remonato (✉)
Department of Mathematical Sciences, NTNU, Trondheim, Norway

Department of Mathematics, University of Pavia, Pavia, Italy
e-mail: filippo.remonato@ntnu.no

G. Sangalli
Department of Mathematics, University of Pavia, Pavia, Italy

IMATI-CNR "E. Magenes", Pavia, Italy
e-mail: giancarlo.sangalli@unipv.it

extensions, such as non-uniform rational B-splines (NURBS). We refer to the monograph [2] for a detailed description of this approach, as well as to [1] for an overview of the mathematical theory of IGA.

In this work we present three general free boundary algorithms. The first algorithm is an extension of the finite elements approach of [12, 13] to IGA. Since the finite element basis produces meshes with straight edges, the authors needed a workaround to approximate the curvature of the boundary; in the new IGA framework this can be avoided thanks to the natural description of curved geometries through spline functions. IGA of free boundary problems was already considered in [11, 21]; our second algorithm uses and extends these approaches to problems with periodic conditions. Our third and most efficient scheme uses instead an isogeometric variational collocation approach based on the superconvergent points presented in [8, 14]. The choice of applying an IGA collocation method is a novelty in this setting and, moreover, allows for a fast computation of the solution. While speed is marginally important in the benchmarks considered in this work, it becomes a major concern when one needs to address more complicated problems.

All the algorithms are based on shape calculus techniques, see for example [5, 19]. This results in the three algorithms being of quasi-Newton type.

Our interest in free boundary problems is motivated by a separate analysis, in progress at the time of writing, of the periodic solutions of the Euler equations describing the flow of an incompressible fluid over a rigid bottom. The analytical literature on this problem is quite extensive, with results regarding irrotational flows [9], the limiting Stokes waves [20], or waves on a rotational current containing one or multiple critical layers [6, 24]. The numerical experiments so far have used finite differences methods [3], boundary-integral formulations [18], or finite elements [16]. Several other examples and numerical experiments, also based on boundary formulations, can additionally be found in [22].

This paper is organised as follows: In Sect. 2 we describe the details of free boundary problem, and present two weak formulations that will constitute our starting point for the algorithms. In Sect. 3 we first introduce the necessary shape calculus tools, and then proceed to linearise the aforementioned weak forms. This will produce the correct formulations on which to base our quasi-Newton steps. Section 4 describes the discrete spaces used in the numerical schemes along with the structure of the algorithms. Finally, Sect. 5 presents the numerical benchmarks and the results we obtained. We summarise the results and draw our conclusions in Sect. 6.

2 Free Boundary Problem

Let Ω_0 be a domain used as reference configuration with $\partial \Omega_0 = \Gamma_\mathcal{D} \cup \Gamma_\mathcal{P} \cup \Gamma_0$; $\Gamma_\mathcal{D}$ being the (fixed) bottom boundary with Dirichlet data, $\Gamma_\mathcal{P}$ the (fixed) vertical boundary, and Γ_0 the (free) upper part of the boundary. Moreover, let D be a rectangle with basis $\Gamma_\mathcal{D}$ of length λ, containing Ω_0 and all its possible deformations.

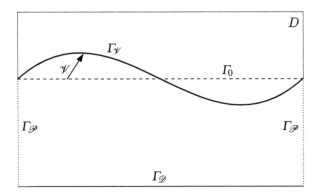

Fig. 1 The setting of our problem. The vector field $\mathcal{V}$ deforms the reference free boundary Γ_0 (dashed line) into the free boundary $\Gamma_\mathcal{V}$ (thick solid line). The lateral boundary is indicated with $\Gamma_\mathcal{P}$, while the thin solid line represents the fixed flat bottom boundary $\Gamma_\mathcal{D}$. The physical domain and its deformations are contained in a larger rectangle D

For M a domain and Γ a curve, we denote with $C^{k,l}(M, \mathbb{R}^2)$ the space of (k, l)-Hölder continuous functions defined on M with values in $\mathbb{R}^2$ and by $C_0^{k,l}(\Gamma, \mathbb{R}^2)$ the subspace of $C^{k,l}(\Gamma, \mathbb{R}^2)$ with compact support, in particular vanishing at the two extremes of the curve. Then, the set of admissible vector fields acting on the reference domain is defined as $\Theta = \{\mathcal{V} \in C^{0,1}(D, \mathbb{R}^2) \cap C_0^{1,1}(\Gamma_0, \mathbb{R}^2) \,|\, \mathcal{V} = 0 \text{ on } \Gamma_\mathcal{D} \text{ and } \mathcal{V}(\cdot, y) \ \lambda\text{-periodic}\}$. We encode the deformation of the upper part of the boundary, Γ_0, as the action of a vector field $\mathcal{V} \in \Theta$ such that the deformed domain is smooth enough, does not have self intersections and does not touch the bottom $\Gamma_\mathcal{D}$. For this reason we denote the deformed free boundary with $\Gamma_\mathcal{V} = \{x \in \mathbb{R}^2 \,|\, x = x_0 + \mathcal{V}(x_0),\ x_0 \in \Gamma_0\}$. Analogously, $\Omega_\mathcal{V}$ will denote the physical domain with boundary $\partial \Omega_\mathcal{V} = \Gamma_\mathcal{D} \cup \Gamma_\mathcal{P} \cup \Gamma_\mathcal{V}$; see Fig. 1 for a representation of this setting. We remark that Γ_0 is in general not flat.

The Bernoulli-type free boundary problem (FBP) we are interested in can then be posed as searching for a pair $(u, \mathcal{V})$, both λ-periodic in the x-direction, such that

$$-\Delta u = f \quad \text{in } \Omega_\mathcal{V} \tag{1a}$$

$$u = h \quad \text{on } \Gamma_\mathcal{V} \cup \Gamma_\mathcal{D} \tag{1b}$$

$$\partial_n u = g \quad \text{on } \Gamma_\mathcal{V} \tag{1c}$$

where $\partial_n u = \nabla u \cdot n$ is the outward normal derivative of u. The functions f, h, and g are defined in D and are compatible with the periodicity requirement. We will consider h and g continuous, with g strictly positive and bounded away from zero.[1]

---

[1] The strict positivity is not strictly necessary: If $g < 0$ one could, for instance, keep track of the sign of g in the numerical method directly. However, g has to have a definite sign everywhere on $\Gamma_\mathcal{V}$.

Remark 1 The analytical treatment of the problem with periodic boundary conditions does not differ much from the case with pure Dirichlet conditions, which we also consider in our numerical benchmarks. In the Dirichlet problems, the pair $(u, \mathcal{V})$ is not required to be λ-periodic in the x- direction and the Dirichlet condition (1b) is imposed on $\Gamma_{\mathcal{V}} \cup \Gamma_{\mathcal{D}} \cup \Gamma_{\mathcal{P}}$.

2.1 Weak Formulation

To obtain a formulation of (1) suitable for a numerical scheme we first follow the steps presented in [12]. This approach leads to two distinct, coupled weak forms. Given the space $H^1_{per}(\Omega_{\mathcal{V}}) := \{u \in H^1(\Omega_{\mathcal{V}}) \mid u(\cdot, y) \ \lambda\text{-periodic}\}$, for a known function r periodic in the x-direction we define the space

$$H^1_{r,\Gamma_{\mathcal{D}}}(\Omega_{\mathcal{V}}) := \{\varphi \in H^1_{per}(\Omega_{\mathcal{V}}) \mid \varphi = r \text{ on } \Gamma_{\mathcal{D}}\}.$$

The first weak form is then obtained using (1a), (1c), and the part of (1b) pertaining to $\Gamma_{\mathcal{D}}$. We select test functions $\varphi \in H^1_{0,\Gamma_{\mathcal{D}}}(\Omega_{\mathcal{V}})$ and apply Green's formula once to obtain

$$\int_{\Omega_{\mathcal{V}}} \nabla u \cdot \nabla \varphi \, d\Omega - \int_{\Gamma_{\mathcal{V}}} g \varphi \, d\Gamma = \int_{\Omega_{\mathcal{V}}} f \varphi \, d\Omega. \tag{2}$$

Using the part of (1b) on $\Gamma_{\mathcal{V}}$ we employ test functions $v \in H^1_{per}(\Gamma_{\mathcal{V}})$ and write the second weak form simply as

$$\int_{\Gamma_{\mathcal{V}}} uv \, d\Gamma = \int_{\Gamma_{\mathcal{V}}} hv \, d\Gamma. \tag{3}$$

We select the trial function space by requiring $u \in H^1_{h,\Gamma_{\mathcal{D}}}(\Omega_{\mathcal{V}})$, thereby strongly imposing the Dirichlet boundary conditions on $\Gamma_{\mathcal{D}}$. This leads to the definition of two linear forms:

$$M_1(u, \mathcal{V}; \varphi) := \int_{\Omega_{\mathcal{V}}} \nabla u \cdot \nabla \varphi \, d\Omega - \int_{\Gamma_{\mathcal{V}}} g \varphi \, d\Gamma - \int_{\Omega_{\mathcal{V}}} f \varphi \, d\Omega, \tag{4}$$

$$M_2(u, \mathcal{V}; v) := \int_{\Gamma_{\mathcal{V}}} uv \, d\Gamma - \int_{\Gamma_{\mathcal{V}}} hv \, d\Gamma. \tag{5}$$

Thus, with this approach the problem is defined as:
Search for $(u, \mathcal{V}) \in H^1_{h,\Gamma_{\mathcal{D}}}(\Omega_{\mathcal{V}}) \times \Theta$ such that

$$M_1(u, \mathcal{V}; \varphi) = 0,$$
$$M_2(u, \mathcal{V}; v) = 0,$$

for all test functions $(\varphi, v) \in H^1_{0,\Gamma_{\mathcal{D}}}(\Omega_{\mathcal{V}}) \times H^1_{per}(\Gamma_{\mathcal{V}})$.

2.2 Very-Weak Formulation

We now follow the approach of [21]. The main difference from the previous formulation is that we write a single very-weak formulation containing information from all boundary conditions.

Considering the subspace $H^2_{0,\Gamma_D}(\Omega_\mathcal{V}) := \{\varphi \in H^1_{0,\Gamma_D}(\Omega_\mathcal{V}) \,|\, \varphi \in H^2(\Omega_\mathcal{V})\}$, we multiply (1a) by a test function $\varphi \in H^2_{0,\Gamma_D}(\Omega_\mathcal{V})$; integrating by parts twice leads to

$$-\int_{\Omega_\mathcal{V}} (u-h)\,\Delta\varphi\,d\Omega + \int_{\Omega_\mathcal{V}} \nabla h \cdot \nabla\varphi\,d\Omega = \int_{\Omega_\mathcal{V}} f\,\varphi\,d\Omega + \int_{\Gamma_\mathcal{V}} \varphi\,g\,d\Gamma, \quad (6)$$

which we demand to be satisfied for all $\varphi \in H^2_{0,\Gamma_D}(\Omega_\mathcal{V})$. In view of the above formulation we can then select the trial function space simply as $H^1_{per}(\Omega_\mathcal{V})$. The Dirichlet boundary conditions are therefore all imposed weakly.

From Equation (6) we define the linear form

$$N(u,\mathcal{V};\varphi) := -\int_{\Omega_\mathcal{V}} (u-h)\,\Delta\varphi\,d\Omega + \int_{\Omega_\mathcal{V}} \nabla h \cdot \nabla\varphi\,d\Omega$$
$$- \int_{\Omega_\mathcal{V}} f\,\varphi\,d\Omega - \int_{\Gamma_\mathcal{V}} \varphi\,g\,d\Gamma. \quad (7)$$

Thus, with this approach the problem is defined as:
Search for $(u,\mathcal{V}) \in H^1_{per}(\Omega_\mathcal{V}) \times \Theta$ such that

$$N(u,\mathcal{V};\varphi) = 0$$

for all test functions $\varphi \in H^2_{0,\Gamma_D}(\Omega_\mathcal{V})$.

Note that this very-weak formulation cannot be used directly to implement a numerical scheme, as the trial and test spaces are unbalanced.

3 Linearising the FBP

We now proceed in deriving a quasi-Newton algorithm to solve the free boundary problem. The dependence on the geometry of the domain is handled through shape calculus techniques to express the derivatives with respect to the vector field $\mathcal{V}$.

3.1 Shape Derivatives

Here we briefly state the shape calculus results we will need for the linearisation. An in-depth analysis of the assumptions and regularity requirements can be found in the original work by Delfour, Zolésio, and Sokolowski [5, 19]. An overview of shape calculus presented with a more modern approach can also be found in [11].

Let O be a family of admissible (smooth enough) domains; a functional J is called a *shape functional* if $J : O \to \mathbb{R}$. Note therefore that for a fixed function u and test functions φ and v, the maps defined by the linear forms introduced earlier are shape functionals provided we identify each element $\mathcal{V} \in \Theta$ with the domain $\Omega_\mathcal{V}$ in which Ω_0 is deformed by the action of $\mathcal{V}$.

In the particular cases of a domain functional $J(\mathcal{V}) = \int_{\Omega_\mathcal{V}} \psi \, d\Omega$ and a boundary functional $F(\mathcal{V}) = \int_{\Gamma_\mathcal{V}} \phi \, d\Gamma$, with ψ and ϕ smooth functions in $\mathbb{R}^2$ independent of $\mathcal{V}$, the shape derivatives of J and F are described by the following *Hadamard formulas*:

$$\langle \partial_\mathcal{V} J(\mathcal{V}), \delta\mathcal{V} \rangle = \int_{\Gamma_\mathcal{V}} \psi \, \delta\mathcal{V} \cdot n \, d\Gamma \tag{8a}$$

$$\langle \partial_\mathcal{V} F(\mathcal{V}), \delta\mathcal{V} \rangle = \int_{\Gamma_\mathcal{V}} \left(\partial_n \phi + H\phi \right) \delta\mathcal{V} \cdot n \, d\Gamma \tag{8b}$$

where $\delta\mathcal{V} \in \Theta$ is a perturbation of the vector field, H is the *signed (additive) curvature* of $\Gamma_\mathcal{V}$ and n is the normal vector pointing outward. In particular, considering a parametrization of the free boundary $\Gamma_\mathcal{V}$ defined as $\gamma(t) = (t, \eta(t))$, then

$$H := -\frac{\eta''}{\left[1 + (\eta')^2\right]^{3/2}}.$$

3.2 Linearisation of the Weak Formulation

Let us first consider the linear forms (4) and (5). We want to linearise M_1 and M_2 with respect to u and $\mathcal{V}$ at an arbitrary approximated solution $(u^*, \mathcal{V}^*) \in H^1_{h,\Gamma_\mathcal{D}}(\Omega_{\mathcal{V}^*}) \times \Theta$.

Since the dependence of M_1 and M_2 on u is affine, their Gâteaux derivatives with respect to u in the direction $\delta u \in H^1_{0,\Gamma_\mathcal{D}}(\Omega_{\mathcal{V}^*})$ are simply given by:

$$\langle \partial_u M_1[u^*, \mathcal{V}^*; \varphi], \delta u \rangle = \int_{\Omega_{\mathcal{V}^*}} \nabla \delta u \cdot \nabla \varphi \, d\Omega \tag{9a}$$

$$\langle \partial_u M_2[u^*, \mathcal{V}^*; v], \delta u \rangle = \int_{\Gamma_{\mathcal{V}^*}} \delta u \, v \, d\Gamma. \tag{9b}$$

The linearisation with respect to the vector field $\mathcal{V}$ in the direction $\delta\mathcal{V} \in \Theta$ is performed using the Hadamard formulas; we obtain:

$$\langle \partial_\mathcal{V} M_1[u^*, \mathcal{V}^*; \varphi], \delta\mathcal{V} \rangle = \int_{\Gamma_{\mathcal{V}^*}} \nabla u^* \cdot \nabla\varphi \, \delta\mathcal{V} \cdot n \, d\Gamma$$

$$- \int_{\Gamma_{\mathcal{V}^*}} \left[K_H \varphi + g \, \partial_n \varphi \right] \delta\mathcal{V} \cdot n \, d\Gamma \quad (10a)$$

$$\langle \partial_\mathcal{V} M_2[u^*, \mathcal{V}^*; v], \delta\mathcal{V} \rangle = \int_{\Gamma_{\mathcal{V}^*}} \left(\partial_n u^* - \partial_n h + H(u^* - h) \right) v \, \delta\mathcal{V} \cdot n \, d\Gamma$$

$$+ \int_{\Gamma_{\mathcal{V}^*}} (u^* - h) \, \partial_n v \, \delta\mathcal{V} \cdot n \, d\Gamma \quad (10b)$$

where $K_H := \partial_n g + H g + f$, and H is the curvature of $\Gamma_{\mathcal{V}^*}$.

A Newton step at the point $(u^*, \mathcal{V}^*)$ has then the following structure: Search for $\delta u \in H^1_{0, \Gamma_D}(\Omega_{\mathcal{V}^*})$ and $\delta\mathcal{V} \in \Theta$ such that

$$\langle \partial_u M_1[u^*, \mathcal{V}^*; \varphi], \delta u \rangle + \langle \partial_\mathcal{V} M_1[u^*, \mathcal{V}^*; \varphi], \delta\mathcal{V} \rangle = -M_1(u^*, \mathcal{V}^*; \varphi) \quad (11a)$$

$$\langle \partial_u M_2[u^*, \mathcal{V}^*; v], \delta u \rangle + \langle \partial_\mathcal{V} M_2[u^*, \mathcal{V}^*; v], \delta\mathcal{V} \rangle = -M_2(u^*, \mathcal{V}^*; v) \quad (11b)$$

for all $(\varphi, v) \in H^1_{0, \Gamma_D}(\Omega_{\mathcal{V}^*}) \times H^1_{per}(\Gamma_{\mathcal{V}^*})$.

Therefore, summing all the contributions, we search for $\tilde{u} = u^* + \delta u \in H^1_{h, \Gamma_D}(\Omega_{\mathcal{V}^*})$ and $\delta\mathcal{V} \in \Theta$ such that

$$\int_{\Omega_{\mathcal{V}^*}} \nabla\tilde{u} \cdot \nabla\varphi \, d\Omega + \int_{\Gamma_{\mathcal{V}^*}} (\partial_n u^* - g) \partial_n \varphi \, \delta\mathcal{V} \cdot n \, d\Gamma + \int_{\Gamma_{\mathcal{V}^*}} \nabla_\Gamma u^* \cdot \nabla\varphi \, \delta\mathcal{V} \cdot n \, d\Gamma$$

$$- \int_{\Gamma_{\mathcal{V}^*}} K_H \varphi \, \delta\mathcal{V} \cdot n \, d\Gamma = \int_{\Omega_{\mathcal{V}^*}} f \varphi \, d\Omega + \int_{\Gamma_{\mathcal{V}^*}} g \varphi \, d\Gamma \quad (12a)$$

$$\int_{\Gamma_{\mathcal{V}^*}} \tilde{u} \, v \, d\Gamma + \int_{\Gamma_{\mathcal{V}^*}} \left[\left(\partial_n u^* - \partial_n h + H(u^* - h) \right) v + (u^* - h) \partial_n v \right] \delta\mathcal{V} \cdot n \, d\Gamma$$

$$= \int_{\Gamma_{\mathcal{V}^*}} h \, v \, d\Gamma \quad (12b)$$

for all $\varphi \in H^1_{0, \Gamma_D}(\Omega_{\mathcal{V}^*})$ and $v \in H^1_{per}(\Gamma_{\mathcal{V}^*})$.

In the above steps we used the *tangential gradient splitting*, with the tangential gradient of a real function being defined as $\nabla_\Gamma(\cdot) = \nabla(\cdot) - \partial_n(\cdot) n$.

So far we carried out the computations in full generality, and (12) is an exact Newton scheme. We now proceed to comment on, and apply, some simplifications.

Simplification 1 Without loss of generality one can consider $\partial_n h = 0$ on $\Gamma_{\mathcal{V}^*}$. Furthermore, we consider the case of constant data $h = h_0$, so then $\nabla_\Gamma h = 0$ and $\nabla h = 0$ on $\Gamma_{\mathcal{V}^*}$. □

Simplification 2 The above formulas can be simplified further by considering, on $\Gamma_{\mathcal{V}^*}$, $u^* = h_0$ and $\partial_n u^* = g$. These conditions are consistent with the *exact solution* of the FBP, and lead to a quasi-Newton method as in [12, 21]. □

Applying the above simplifications produces the following quasi-Newton scheme: Search for $\tilde{u} \in H^1_{h,\Gamma_D}(\Omega_{\mathcal{V}^*})$ and $\delta\mathcal{V} \in \Theta$ such that

$$\int_{\Omega_{\mathcal{V}^*}} \nabla \tilde{u} \cdot \nabla \varphi \, d\Omega - \int_{\Gamma_{\mathcal{V}^*}} K_H \varphi \, \delta\mathcal{V} \cdot n \, d\Gamma = \int_{\Omega_{\mathcal{V}^*}} f \varphi \, d\Omega + \int_{\Gamma_{\mathcal{V}^*}} g \varphi \, d\Gamma \quad (13a)$$

$$\int_{\Gamma_{\mathcal{V}^*}} \tilde{u} \, v \, d\Gamma + \int_{\Gamma_{\mathcal{V}^*}} g \, v \, \delta\mathcal{V} \cdot n \, d\Gamma = \int_{\Gamma_{\mathcal{V}^*}} h_0 \, v \, d\Gamma \quad (13b)$$

for all $(\varphi, v) \in H^1_{0,\Gamma_D}(\Omega_{\mathcal{V}^*}) \times H^1_{per}(\Gamma_{\mathcal{V}^*})$.

Remark 2 The Simplification 2 above is the reason why the scheme (13) is not an exact Newton scheme, but only quasi-Newton method: The derivatives are not calculated in the current approximation, but rather they are an approximation of the derivatives at the exact solution. This has the consequence that (13) does not achieve quadratic convergence, as we will see in Sect. 5.

3.3 Linearisation of the Very-Weak Formulation

We now want to derive a linearisation for (7) at an arbitrary approximated solution $(u^*, \mathcal{V}^*)$, where as before $u^* \in H^1_{per}(\Omega_{\mathcal{V}^*})$ and $\mathcal{V}^* \in \Theta$. The Gâteaux derivative of N at $(u^*, \mathcal{V}^*)$ with respect to u in the direction δu is given by

$$\langle \partial_u N[u^*, \mathcal{V}^*; \varphi], \delta u \rangle = -\int_{\Omega_{\mathcal{V}^*}} \delta u \, \Delta \varphi \, d\Omega. \quad (14)$$

The linearisation with respect to the vector field is again performed using the Hadamard formulas (8):

$$\langle \partial_\mathcal{V} N[u^*, \mathcal{V}^*; \varphi], \delta\mathcal{V} \rangle = \int_{\Gamma_{\mathcal{V}^*}} \nabla h \cdot \nabla \varphi \, \delta\mathcal{V} \cdot n \, d\Gamma - \int_{\Gamma_{\mathcal{V}^*}} (u^* - h) \Delta \varphi \, \delta\mathcal{V} \cdot n \, d\Gamma$$

$$- \int_{\Gamma_{\mathcal{V}^*}} \left[K_H \varphi + g \, \partial_n \varphi \right] \delta\mathcal{V} \cdot n \, d\Gamma. \quad (15)$$

A Newton step at the point $(u^*, \mathcal{V}^*)$ has then the following form: Search for $\delta u \in H^1_{0,\Gamma_D}(\Omega_{\mathcal{V}^*})$ and $\delta \mathcal{V} \in \Theta$ such that

$$\langle \partial_u N[u^*, \mathcal{V}^*; \varphi], \delta u \rangle + \langle \partial_{\mathcal{V}} N[u^*, \mathcal{V}^*; \varphi], \delta \mathcal{V} \rangle = -N(u^*, \mathcal{V}^*; \varphi), \quad (16)$$

for all $\varphi \in H^2_{0,\Gamma_D}(\Omega_{\mathcal{V}})$.

Summing the various terms we then search for $\tilde{u} = u^* + \delta u \in H^1_{h,\Gamma_D}(\Omega_{\mathcal{V}^*})$ and $\delta \mathcal{V} \in \Theta$ such that

$$\int_{\Omega_{\mathcal{V}^*}} (h - \tilde{u}) \Delta \varphi \, d\Omega - \int_{\Gamma_{\mathcal{V}^*}} \left[K_H \varphi + g \, \partial_n \varphi + (u^* - h) \Delta \varphi \right] \delta \mathcal{V} \cdot n \, d\Gamma$$
$$+ \int_{\Gamma_{\mathcal{V}^*}} \nabla h \cdot \nabla \varphi \, \delta \mathcal{V} \cdot n \, d\Gamma = \int_{\Gamma_{\mathcal{V}^*}} g \varphi \, d\Gamma + \int_{\Omega_{\mathcal{V}^*}} f \varphi \, d\Omega - \int_{\Omega_{\mathcal{V}^*}} \nabla h \cdot \nabla \varphi \, d\Omega, \quad (17)$$

for all $\varphi \in H^2_{0,\Gamma_D}(\Omega_{\mathcal{V}})$.

Upon application of Simplifications 1 and 2, we obtain the following quasi-Newton scheme: Search for $\tilde{u} \in H^1_{h,\Gamma_D}(\Omega_{\mathcal{V}^*})$ and $\delta \mathcal{V} \in \Theta$ such that

$$\int_{\Omega_{\mathcal{V}^*}} (h - \tilde{u}) \Delta \varphi \, d\Omega - \int_{\Gamma_{\mathcal{V}^*}} \left[K_H \varphi + g \, \partial_n \varphi \right] \delta \mathcal{V} \cdot n \, d\Gamma$$
$$= \int_{\Gamma_{\mathcal{V}^*}} g \varphi \, d\Gamma + \int_{\Omega_{\mathcal{V}^*}} f \varphi \, d\Omega - \int_{\Omega_{\mathcal{V}^*}} \nabla h \cdot \nabla \varphi \, d\Omega, \quad (18)$$

for all $\varphi \in H^2_{0,\Gamma_D}(\Omega_{\mathcal{V}})$.

As we pointed out above, we cannot yet employ this formulation to produce a numerical scheme; we need to extract the strong form implied by (18) and then write a new weak formulation. Using standard variational arguments one can see that such strong form is:

$$-\Delta \tilde{u} = f \quad \text{in } \Omega_{\mathcal{V}^*} \quad (19a)$$
$$\partial \tilde{u}_n - K_H \, \delta \mathcal{V} \cdot n = g \quad \text{on } \Gamma_{\mathcal{V}^*} \quad (19b)$$
$$\tilde{u} = h \quad \text{on } \Gamma_D \quad (19c)$$
$$g \, \delta \mathcal{V} \cdot n = h_0 - \tilde{u} \quad \text{on } \Gamma_{\mathcal{V}^*}. \quad (19d)$$

Thanks to the initial requirement on g not vanishing, one can solve (19d) for $\delta \mathcal{V} \cdot n$, obtaining the boundary update formula

$$\delta \mathcal{V} \cdot n = \frac{h_0 - \tilde{u}}{g}. \quad (20)$$

Substituting in (19b) and using (19a)–(19c) allows to write the new weak formulation: Search for $\tilde{u} \in H^1_{h,\Gamma_\mathcal{D}}(\Omega_{\mathcal{V}^*})$ such that

$$\int_{\Omega_{\mathcal{V}^*}} \nabla \tilde{u} \cdot \nabla \varphi \, d\Omega - \int_{\Gamma_{\mathcal{V}^*}} \left(K_H \frac{h_0 - \tilde{u}}{g} + g \right) \varphi \, d\Gamma = \int_{\Omega_{\mathcal{V}^*}} f\varphi \, d\Omega, \qquad (21)$$

for all $\varphi \in H^1_{0,\Gamma_\mathcal{D}}(\Omega_{\mathcal{V}^*})$.

Remark 3 Solving Equation (13b) for $\delta\mathcal{V} \cdot n$ one obtains exactly Equation (20). Plugging then into Equation (13b) gives Equation (21). This shows that the two methods, the coupled system (13) and the formulation (21) with boundary update as in (20), are variationally equivalent, so we can expect the behaviours of these two approaches to be very similar. On the other hand, even though they are equivalent in an infinite-dimensional setting, the difference in the way the vector field is handled (as a coupled projection in the former case, or a splitting method in the latter case) may be reflected in the performances at the discretised level. This will indeed be the case, as our numerical tests illustrate.

The strong form (19) will also be used in the implementation of a collocation scheme, outlined in the next section. In passing, we comment that in the case of non-constant Dirichlet data on the free boundary, from Equation (17) one could split the gradient of h into the third integral in its tangential and normal component, and apply the tangential Green's identity [5, p. 367]. See also [21] for details.

4 Numerical Schemes

In our numerical tests we used two Galerkin methods, one arising from (13) and one from (21). The main difference between them is that from the former one obtains a coupled method, while the latter yields a decoupled splitting method. Moreover, we implemented a collocation method to solve the strong form (19).

4.1 B-Splines Based Isogeometric Analysis

This section presents the essentials of B-splines. For more details we refer the interested reader to any of the specialised books on the subject, for instance [7].

A *knot vector* is a set of non-decreasing points $\Xi = \{\xi_1 \leq \ldots \leq \xi_{m+p+1}\}$ with $\xi_i \in \mathbb{R}$ and m the number of basis functions of degree p to be built.

A knot vector is said to be *open* if its first and last knots have multiplicity $p + 1$, and in this case it is customary to take $\xi_1 = 0$ and $\xi_{m+p+1} = 1$. The maximum multiplicity of each internal knot can never exceed p. A knot vector is said to be *uniform* if the knots are equispaced; in this case it is common to take $\xi_1 = -p\tau$ and $\xi_{m+p+1} = p\tau$, with τ the distance between two consecutive knots.

Univariate B-splines functions can be defined using the Cox-de Boor recursion formulas [4] as follows, for $i = 1, \ldots, m$:

for $p = 0$:

$$\hat{\psi}_{i,0}(\xi) := \begin{cases} 1 & \xi_i \leq \xi < \xi_{i+1}, \\ 0 & \text{otherwise,} \end{cases}$$

for $p \geq 1$:

$$\hat{\psi}_{i,p}(\xi) := \begin{cases} \dfrac{\xi - \xi_i}{\xi_{i+p} - \xi_i} \hat{\psi}_{i,p-1}(\xi) + \dfrac{\xi_{i+p+1} - \xi}{\xi_{i+p+1} - \xi_{i+1}} \hat{\psi}_{i+1,p-1}(\xi) & \xi_i \leq \xi < \xi_{i+p+1}, \\ 0 & \text{otherwise,} \end{cases}$$

where we adopt the convention $0/0 = 0$. A B-spline basis function is therefore a piecewise polynomial in every knot span and at the knots it achieves regularity C^{p-l} where l is the multiplicity of the knot. We will always use internal knots of multiplicity one, in order to have maximal regularity.

We denote with $\hat{S}^p := \text{span}\{\hat{\psi}_{i,p} \mid i = 1, \ldots, m\}$ the space spanned by m B-splines of degree p. We will often omit to explicitly indicate the polynomial degree. On a uniform knot vector one can in addition construct a *periodic* basis by appropriately identifying together functions laying at the beginning and at the end of the parametric domain:

$$\hat{S}^p_{per} := \text{span}\{\hat{\psi}^{per}_k\} \quad \text{with} \quad \begin{cases} \hat{\psi}^{per}_k := \hat{\psi}_k + \hat{\psi}_{m-p+k}, & k = 1, \ldots, p; \\ \hat{\psi}^{per}_k := \hat{\psi}_k, & \text{otherwise} \end{cases} \quad (22)$$

Note that $\dim(\hat{S}^p_{per}) = m - p$. Figure 2b shows an example of maximum-regularity periodic B-splines basis with degree $p = 3$.

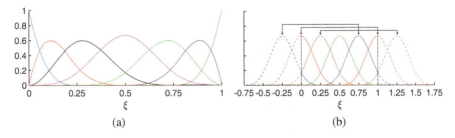

Fig. 2 Example of open and periodic B-spline basis. (**a**) Cubic basis on an open knot vector. (**b**) A periodic cubic basis on a uniform knot vector

We can derive bivariate B-splines spaces, which we indicate in boldface, simply considering the tensor product of univariate ones. Moreover, in our numerical tests we will use the same degree in each parametric direction.

Now, let $\mathbf{F} : \hat{\Omega} \to \Omega$ be a B-spline parametrisation (periodic in the x-direction) of the physical domain Ω, and let $\hat{\mathbf{S}}^p$ be a space spanned by N bivariate B-splines $\hat{\phi}_k$ defined on the parametric domain $\hat{\Omega}$. Then, the corresponding space on Ω is defined as $\mathbf{S}^p := \text{span}\{\phi_k \mid \phi_k = \hat{\phi}_k \circ \mathbf{F}^{-1}, \ k = 1, \ldots, N\}$. We moreover need to introduce a bivariate spline space spanned by functions periodic in x, that we denote $\mathbf{S}^p_{per}$. This space is defined as the push-forward through the geometrical map $\mathbf{F}$ of the cross product between the periodic space $\hat{S}^p_{per}$, and the space $\hat{S}^p$ built from an open knot vector. Note that $\dim(\mathbf{S}^p_{per}) = m(m-p)$.

4.2 Isogeometric Galerkin Methods

In both Galerkin-based schemes we choose as a trial space for $\tilde{u}$

$$\mathbf{V}^p_h := \mathbf{S}^p_{per} \cap H^1_{h,\Gamma_{\mathcal{D}}}(\Omega_{\mathcal{V}*}), \tag{23}$$

while as test space

$$\mathbf{V}^p_0 := \mathbf{S}^p_{per} \cap H^1_{0,\Gamma_{\mathcal{D}}}(\Omega_{\mathcal{V}*}). \tag{24}$$

Note that $\dim(\mathbf{V}^p_h) = \dim(\mathbf{V}^p_0) = (m-p)(m-1)$.

The structure of the two algorithms is illustrated below.

Algorithm 1 - Coupled Galerkin scheme

1: Choose the initial $\mathcal{V}_0$,
2: Given $\mathcal{V}_k$, compute $(\tilde{u}_k, \delta \mathcal{V} \cdot n_k)$ solution of (13) in the domain $\Omega_{\mathcal{V}_k}$,
3: Update the free boundary with $\mathcal{V}_{k+1} = \mathcal{V}_k + (\delta \mathcal{V} \cdot n_k)\mu_k$,
4: Repeat steps 2–3 until $\|(\delta \mathcal{V} \cdot n_k)\mu_k\| \leq \text{tol}$.

Algorithm 2 - Decoupled (splitting) Galerkin scheme

1: Choose the initial $\mathcal{V}_0$,
2: Given $\mathcal{V}_k$, compute $\tilde{u}_k$ solution of (21) in the domain $\Omega_{\mathcal{V}_k}$,
3: Compute $\delta \mathcal{V} \cdot n_k$ from (20),
4: Update the free boundary with $\mathcal{V}_{k+1} = \mathcal{V}_k + (\delta \mathcal{V} \cdot n_k)\mu_k$,
5: Repeat steps 2–4 until $\|(\delta \mathcal{V} \cdot n_k)\mu_k\| \leq \text{tol}$.

The vector field $n_k : \Gamma_{\mathcal{V}_k} \to \mathbb{R}^2$ represents the outward normal derivative to $\Omega_{\mathcal{V}_k}$, while the vector field $\mu_k : \Gamma_{\mathcal{V}_k} \to \mathbb{R}^2$ represents the direction in which the update of the free boundary is performed, and has to satisfy $\mu_k \cdot n_k = 1$. In our tests we choose to perform a vertical update, therefore selecting $\mu_k = \begin{bmatrix} 0, & 1/(n_k)_y \end{bmatrix}$. This choice allows to consider as unknown $\delta \mathcal{V} \cdot n$ instead of $\delta \mathcal{V}$, which permits to discretise (13b) and (20) directly, using S_{per}^p as both the test and trial space. A choice of $\mu_k = n_k$ in the algorithms would instead amount to performing the update in the direction normal to the boundary.

Remark 4 When performing the update by Equation (20) one has to divide two spline functions. The resulting function is therefore, in general, not a spline, and a projection onto the appropriate spline space is required: in our tests we perform an L^2 projection into the space defined by the boundary test functions. After each boundary update, the internal mesh is then fitted; this operation is trivial in our setting because of the shape of the computational domain and because of the vertical update of the free-boundary. Since we only consider moderate deformations, a simple Coons interpolation is used to construct the geometry parametrization. For more details about the numerical algorithms to compute Coons interpolations see e.g. [15].

In Step 4 of Algorithm 1 and Step 5 of Algorithm 2, we use as $\|\cdot\|$ the Euclidean norm of the control point vector of the spline projection of $(\delta \mathcal{V} \cdot n_k)\mu_k$.

4.3 Isogeometric Collocation Method

The isogeometric collocation method presented here is built from (19): We solve (19d) for $\delta \mathcal{V} \cdot n$ and replace its value in (19b), obtaining the following:

$$-\Delta \tilde{u} = f \qquad \text{in } \Omega, \tag{25a}$$

$$\nabla \tilde{u} \cdot n - K_H \frac{h_0 - \tilde{u}}{g} = g \qquad \text{on } \Gamma_\mathcal{V}, \tag{25b}$$

$$\tilde{u} = h \qquad \text{on } \Gamma_\mathcal{D}, \tag{25c}$$

$$\delta \mathcal{V} \cdot n = \frac{h_0 - \tilde{u}}{g} \qquad \text{on } \Gamma_\mathcal{V}. \tag{25d}$$

The structure of this algorithm is summarised below.

Algorithm 3 - Collocation scheme

1: Choose the initial $\mathcal{V}_0$,
2: Given $\mathcal{V}_k$, compute $\tilde{u}_k$, collocated solution of (25a)–(25c),
3: Compute $\delta \mathcal{V} \cdot n_k$ from (25d),
4: Update the free boundary with $\mathcal{V}^{(k+1)} = \mathcal{V}^{(k)} + (\delta \mathcal{V} \cdot n_k)\mu_k$,
5: Repeat steps 2–4 until $\|(\delta \mathcal{V} \cdot n_k)\mu_k\| \leq$ tol.

The vector fields n_k and μ_k and the norm present in Step 5 of Algorithm 3 are defined as in Sect. 4.3.

The solution of (25a)–(25c) and the boundary update (25d) are performed using a collocation approach.

Let us first focus on (25a)–(25c): given the finite dimensional space $\mathbf{V}_h^p$ in which we search for a solution $\tilde{u}$ the idea is to accurately choose a number of points $\tau_1, \ldots, \tau_{(m-1)(m-p)} \in \Omega$, called *collocation points*, and enforce the equations to hold strongly at those points. We recall that $(m-1)(m-p)$ is the dimension of $\mathbf{V}_h^p$, that is, the number of degrees of freedom of the problem. For this formulation, we need basis functions of continuity at least C^2 at each collocation points: we require $p \geq 3$.

The appropriate selection of collocation points is crucial for the rate of convergence. Most of the classical choices of collocation points, for example, return suboptimal convergence rate even in a Poisson problem, contrary to the Galerkin approach which is optimal [17]. However, the recent work [14] suggests the use of a particular subset of Galerkin-superconvergent points, called clustered superconvergent points (CSP), as collocation points. This choice, that is the one that we adopt here, succeeds in achieving optimality for at least odd degrees B-splines discretisations. In particular, the collocation points we use for the periodic problem (25) are obtained by taking the cross product between the $m - p$ univariate periodic CSP and the m univariate Dirichlet CSP (see [14] for more details). In our tests we included also problems with only Dirichlet boundary conditions. In that case the collocation points are selected as the push-forward of the cross-product of the univariate Dirichlet CSP points in the two parametric directions. Figure 3 shows an example of CSP points in both the parametric and physical domain. Note that we do not take any collocation points on the boundary $\{(x, 0) \mid 0 \leqslant x \leqslant 1\}$, because we enforce the Dirichlet boundary conditions in the finite dimensional space that we consider, cf. (23).

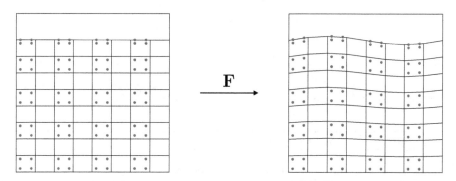

Fig. 3 CSP collocation points in the parametric and in the physical domain. The points are the cross product of the periodic CSP points in the x-direction and the Dirichlet CSP points in the y-direction

Similarly, the free boundary update is performed searching for $\delta V \cdot n \in S_{per}^p$ such that equation (25d) is satisfied in the $m-p$ univariate periodic CSP belonging to Γ_V.

The whole collocation procedure results in a fully-collocated scheme for problem (25).

5 Numerical Results

This section collects our numerical results. All algorithms have been implemented in Matlab using the GeoPDEs suite. GeoPDEs is an Octave/Matlab software package for isogeometric analysis of partial differential equations [23]. We applied the above Algorithm 1, Algorithm 2, and Algorithm 3 to different types of problems with either Dirichlet or periodic boundary conditions on the vertical sides. We set the tolerance "tol" of Step 4 in Algorithm 1 and of Step 5 in Algorithm 2 and Algorithm 3 equal to 10^{-13}. It is clear that the error quantities in the problems are driven by the position of the free boundary: If the computed boundary matches the exact boundary solution, then the error on the internal function u is simply the standard finite elements (IGA) or collocation approximation error. For this reason, when evaluating the performance of the algorithms we have chosen the error quantities of interest to be the *Dirichlet error*, $\|\tilde{u}(\Gamma_V) - h_0\|_{L^2}$, the error the computed function u commits in satisfying the Dirichlet condition on the free boundary, and the *surface position error*, $\|\Gamma_V - \Gamma_{ex}\|_{L^2}$, the error in the position of the computed free surface.

In the following tests we focus on splines of degree $p=3$ and $p=5$. This is motivated by the fact that, as already mentioned in Sect. 4.3, the CSP-based collocation method achieves an optimal order of convergence only for odd degrees of basis functions, see [14].

5.1 Test 1: Parabolic Boundary, Dirichlet b.c.

This problem is constructed from the exact solution

$$u_{ex}(x, y) = \frac{y}{1+\alpha(x)} + \alpha(x) \frac{y}{1+\alpha(x)} \left(1 - \frac{y}{1+\alpha(x)}\right) \qquad (26)$$

with

$$\alpha(x) = \frac{1}{4} x (1-x).$$

The solution u_{ex} attains constant value $u_{ex}|_{\Gamma_\mathcal{V}} = 1$ on the *parabolic curve* $\Gamma_{ex} = \{(x, 1+\alpha(x)) \mid 0 \leq x \leq 1\}$, which is therefore the exact free-boundary solution of the problem.

The data for problem (1) are then found as follows:

$$f = -\Delta u_{ex},$$

$$g = \nabla u_{ex} \cdot \left(\tfrac{1}{2}x - \tfrac{1}{4}, 1\right) \Big/ \sqrt{1 + \left(\tfrac{1}{2}x - \tfrac{1}{4}\right)^2}.$$

We cast this problem with complete Dirichlet boundary conditions without requiring any periodicity in the x-direction, see also Remark 1. This amounts to imposing in (1b) $u = h$ on $\Gamma_\mathcal{V} \cup \Gamma_\mathcal{D} \cup \Gamma_\mathcal{P}$ with $h = h_0 = 1$ on $\Gamma_\mathcal{V}$ and $h = y$ on $\Gamma_\mathcal{D} \cup \Gamma_\mathcal{P}$. We start our algorithms with $\Gamma_0 = \{(x, 1) \mid 0 \leq x \leq 1\}$ as an initial guess for the boundary. Note that this is the same setting as the "Testcase I: Parabolic Free-Boundary" presented in [21, Section 5.2].

Figure 4 shows the first three iterations of the boundary update, together with the exact boundary solution, performed with a mesh with only 1 element and cubic basis functions. Those iterations have in particular been performed with Algorithm 2, but Algorithms 1 and 3 yielded identical results.

Figure 5 instead shows a comparison of the three different approaches using cubic basis functions. The error plots show that Algorithm 1 improves the convergence speed once the solution is close enough. The same behaviour is present also in the collocated scheme, Algorithm 3, albeit to a less degree, while it is not that apparent in Algorithm 2. However, the performance of all three algorithms is quite similar on this test problem. Note that, in contrast to [21], we do not see a plateau in the surface error quantities and machine precision is reached for all the tested mesh-sizes.

To provide the reader with more details, in Table 1 we write the numerical values of the Dirichlet errors corresponding to Fig. 5, panel (c), and the order of

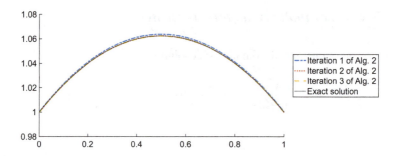

Fig. 4 The first three iterations of the boundary update computed with Algorithm 2, for the Test 1 case, using a one element mesh and cubic basis starting from a flat free boundary guess

Isogeometric Methods for Free Boundary Problems

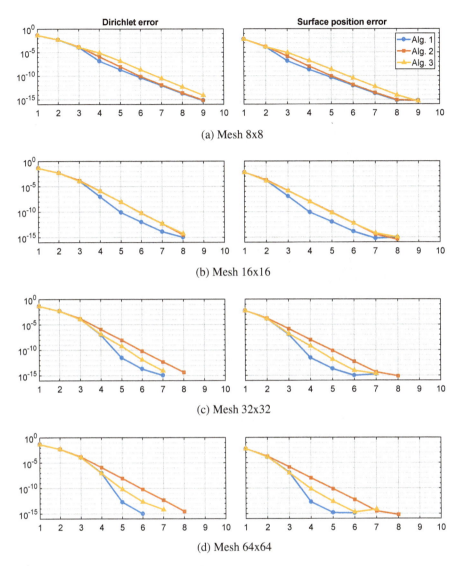

Fig. 5 A comparison of the three algorithms on Test 1 for different mesh sizes with cubic basis functions. In all the subplots, blue circles refer to Algorithm 1, red squares refer to Algorithm 2 and yellow triangles refer to Algorithm 3. (**Left**) The Dirichlet error $\|\tilde{u}(\Gamma_{\mathcal{V}}) - h\|_{L^2}$ as a function of the iterations. (**Right**) The surface position error $\|\Gamma_{\mathcal{V}} - \Gamma_{ex}\|_{L^2}$. (**a**) Mesh 8×8. (**b**) Mesh 16×16. (**c**) Mesh 32×32. (**d**) Mesh 64×64

convergence at each iteration k estimated through the formula

$$order(k) = \frac{\log(err(k)) - \log(err(k+1))}{\log(err(k-1)) - \log(err(k))}, \tag{27}$$

Table 1 Dirichlet errors for the test in Fig. 5 panel (c), i.e. for the mesh with 32 × 32 elements. In parenthesis is the estimated order of convergence, calculated as in Equation (27)

Iteration	Algorithm 1		Algorithm 2		Algorithm 3	
	Error	Order	Error	Order	Error	Order
1	$0.46 \cdot 10^{-1}$		$0.46 \cdot 10^{-1}$		$0.46 \cdot 10^{-1}$	
2	$0.51 \cdot 10^{-2}$	1.73	$0.51 \cdot 10^{-2}$	1.60	$0.51 \cdot 10^{-2}$	1.74
3	$1.09 \cdot 10^{-4}$	1.83	$1.45 \cdot 10^{-4}$	1.35	$1.06 \cdot 10^{-4}$	1.75
4	$9.47 \cdot 10^{-8}$	1.47	$1.17 \cdot 10^{-6}$	1.02	$1.21 \cdot 10^{-7}$	0.80
5	$2.78 \cdot 10^{-12}$	0.47	$8.46 \cdot 10^{-9}$	0.99	$5.18 \cdot 10^{-10}$	1.10
6	$1.99 \cdot 10^{-14}$	0.61	$6.20 \cdot 10^{-11}$	0.99	$1.23 \cdot 10^{-12}$	0.87
7	$9.73 \cdot 10^{-16}$		$4.60 \cdot 10^{-13}$	0.90	$6.06 \cdot 10^{-15}$	
8	–		$5.40 \cdot 10^{-15}$		–	

where $err(k)$ is the error at iteration k. We do not get clear evidence on the order of convergence. We observe superlinear convergence at the very first iterations and then a linear behaviour, at least in Algorithm 2 and Algorithm 3.

Algorithm 3 turns out to be more efficient in terms of computational time vs number of iterations, than the two Galerkin approaches, something which is expected of a collocation scheme. However, at the moment, we could not quantify the speedup obtained by using the Algorithm 3 because, for a fair comparison, we would need optimized versions of the codes.

5.2 Test 2: Sinusoidal Boundary, Dirichlet b.c.

We now give an example where a plateau in the error is to be expected, and is actually found. The problem data is derived as for Test 1 with an exact solution given by Equation (26) but with

$$\alpha_{ex}(x) = \frac{1}{16} \sin(2\pi x),$$

so that the exact boundary $\Gamma_{ex} = \left\{(x, 1 + \alpha(x)) \mid 0 \leqslant x \leqslant 1\right\}$ is now a sinusoidal curve. The boundary conditions are maintained of Dirichlet type, imposing in (1b) $u = h$ on $\Gamma_{\mathcal{V}} \cup \Gamma_{\mathcal{D}} \cup \Gamma_{\mathcal{P}}$ with $h = h_0 = 1$ on $\Gamma_{\mathcal{V}}$ and $h = y$ on $\Gamma_{\mathcal{D}} \cup \Gamma_{\mathcal{P}}$. Figure 6 shows the first three boundary updates performed by Algorithm 3. The mesh is made of 8 elements, and the basis is cubic. The initial boundary is again taken as the flat curve $\Gamma_0 = \left\{(x, 1) \mid 0 \leqslant x \leqslant 1\right\}$

Figures 7 and 8 show the error quantities vs iterations for the three algorithms obtained by using a cubic and quintic basis, respectively. Even if we do not report the precise values of the computed orders of convergence, we remark that for all the

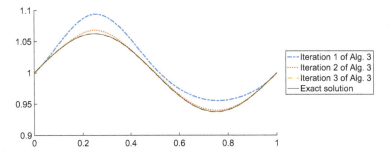

Fig. 6 The first three iterations of the boundary update computed with Algorithm 3, for the Test 2 case, with sinusoidal boundary and Dirichlet conditions, with an 8 elements mesh and cubic basis, starting from a flat free boundary guess

three algorithms the convergence has the same behaviour as in Test 1 case, before reaching the plateau of the error.

As the mesh is refined we note that the collocation algorithm, Algorithm 3, has a slightly higher error than the other two approaches. The surface position error, moreover, is abated with finer meshes in all approaches but remains always present. This is due to the fact that a cubic and quintic B-splines cannot exactly represent a sinusoidal curve, and therefore the exact free boundary solution to this problems lies outside of the trial function space. For equal mesh-size, as expected, the value of the plateau of the surface position error is lower when using splines of higher degree.

Lastly, Figs. 7 and 8 show how closely related Algorithms 1 and 2 are, achieving almost identical performance on this benchmark test.

5.3 Test 3: Sinusoidal Boundary, Periodic b.c.

In our third benchmark we employ the same problem data as in Test 2, but now periodic boundary conditions are placed on the lateral sides instead of Dirichlet ones. In this test case we used the highest-possible regularity for the periodic conditions, meaning that the boundary functions are "glued" together with C^{p-1} continuity.

The introduction of the periodic conditions affects the behaviour of the three quasi-Newton schemes, but not dramatically. As shown in Fig. 9, the algorithms require a couple of extra iterations to reach the tolerance in comparison to the Dirichlet boundary condition case. The convergence of the surface position error is also a bit rougher than in the previous cases. However, the relative performances are not at all affected, and all three algorithms are still comparable. As before Algorithm 1 and Algorithm 2 display essentially equal results. In this test we

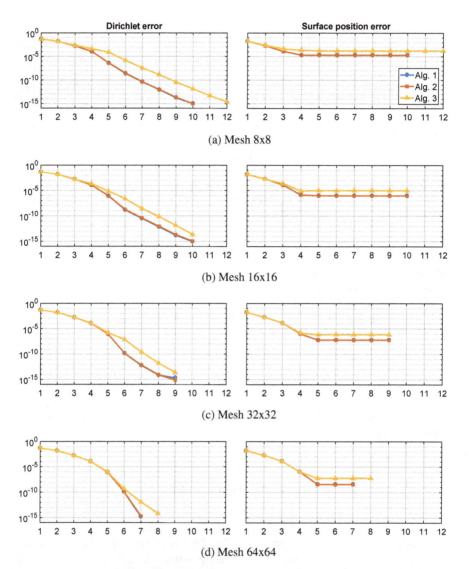

Fig. 7 A comparison of the three algorithms on Test 2 for different mesh sizes with cubic basis functions. In all the subplots, blue circles refer to Algorithm 1, red squares refer to Algorithm 2 and yellow triangles refer to Algorithm 3. (**Left**) The Dirichlet error $\left\|\tilde{u}(\Gamma_\mathcal{V}) - h\right\|_{L^2}$ as a function of the iterations. (**Right**) The surface position error $\left\|\Gamma_\mathcal{V} - \Gamma_{ex}\right\|_{L^2}$. (**a**) Mesh 8×8. (**b**) Mesh 16×16. (**c**) Mesh 32×32. (**d**) Mesh 64×64

Isogeometric Methods for Free Boundary Problems

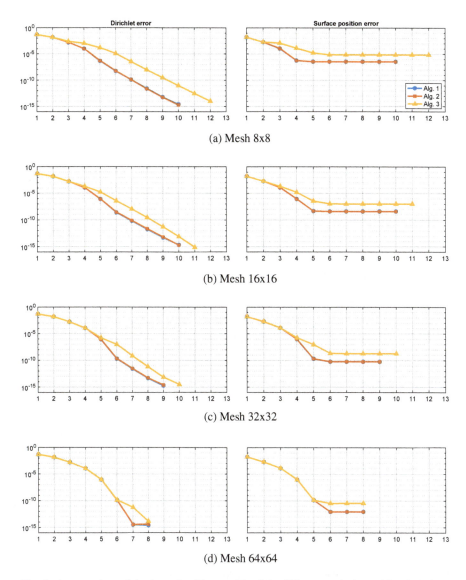

Fig. 8 A comparison of the three algorithms on Test 2 for different mesh sizes with quintic basis functions. In all the subplots, blue circles refer to Algorithm 1, red squares refer to Algorithm 2 and yellow triangles refer to Algorithm 3. (**Left**) The Dirichlet error $\|\tilde{u}(\Gamma_\mathcal{V}) - h\|_{L^2}$ as a function of the iterations. (**Right**) The surface position error $\|\Gamma_\mathcal{V} - \Gamma_{ex}\|_{L^2}$. (**a**) Mesh 8×8. (**b**) Mesh 16×16. (**c**) Mesh 32×32. (**d**) Mesh 64×64

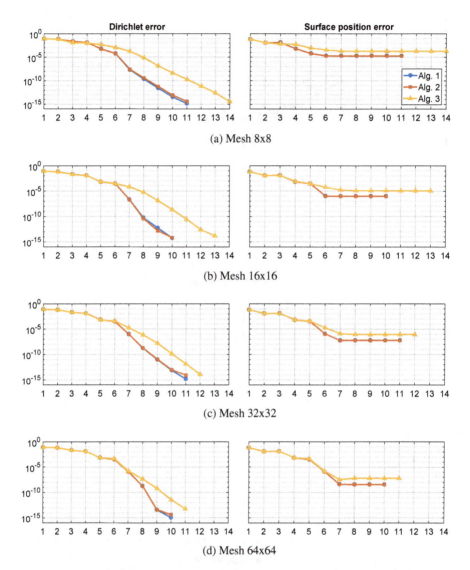

Fig. 9 A comparison of the three algorithms on Test 3 for different mesh sizes with cubic basis functions. In all the subplots, blue circles refer to Algorithm 1, red squares refer to Algorithm 2 and yellow triangles refer to Algorithm 3. (**Left**) The Dirichlet error $\left\|\tilde{u}(\Gamma_\mathcal{V}) - h\right\|_{L^2}$ as a function of the iterations. (**Right**) The surface position error $\left\|\Gamma_\mathcal{V} - \Gamma_{ex}\right\|_{L^2}$. (**a**) Mesh 8×8. (**b**) Mesh 16×16. (**c**) Mesh 32×32. (**d**) Mesh 64×64

kept the same choice for the initial guess for the free boundary: The flat curve $\Gamma_0 = \{(x, 1) | 0 \leqslant x \leqslant 1\}$.

Since the position of the exact free boundary does not lie in the trial functions space formed by the cubic B-splines basis, as in Test 2 a plateau is always reached, even though the level of the plateau is lowered with finer meshes.

Regarding the order of convergence of the methods, in this test case, after a few iterations where the initial error is reduced, the methods exhibit behaviours similar to the one of the other tests.

6 Conclusions

In this work we presented three different isogeometric-based algorithms for free boundary problems: Two follow a Galerkin approach and are an extension or modification of previously existing works, while one is a novel fully collocated scheme. The dependence on the unknown geometry of the domain is handled through shape calculus, which results in a quasi-Newton method to be underlying the update strategy of the free boundary position. The resulting three algorithms yield to a convergence that appears, in general, to be linear and superlinear only in the first iterative steps. While our interests in such algorithms is motivated by future applications, in the present paper we focused on giving a clear description of the implementation and numerical aspects.

We applied and compared the three algorithms to benchmark tests, with either Dirichlet or periodic boundary conditions on the lateral vertical sides of the domain. The results show that, while having slight variations, the performances of all three algorithms are qualitatively comparable, and each of them converged to the correct solution of the problem.

The treatment of free boundary problems is computationally intense, especially in more complex problems. For this reason the efficiency and speed of the algorithm is an important feature that needs to be taken into account. In this respect, even if the collocated algorithm appeared to have slightly worse accuracy and sometimes required one or two extra iterations to reach the convergence tolerance, it proved to significantly outmatch the two Galerkin-based schemes on runtime.

Our future aim is now to apply the algorithms developed here to the resolution of the bifurcation branches of the Euler equations. That problem presents several challenges due to the greater complexity of the equations and the intrinsic non-uniqueness of solutions at the bifurcation points, therefore both efficiency and precision are expected to play an important role.

Acknowledgments The authors would like to thank Rafael Vázquez for the suggestions regarding the implementation of the code. MM and GS were partially supported by the European Research Council through the FP7 Ideas Consolidator Grant *HIGEOM* n.616563 and by the Italian Ministry of Education, University and Research (MIUR) through the "Dipartimenti di Eccellenza Program (2018-2022) - Dept. of Mathematics, University of Pavia". FR was supported by grants no. 231668 and 250070 from the Norwegian Research Council. This support is gratefully acknowledged. MM and GS are members of the INdAM Research group GNCS.

References

1. L. Beirão da Veiga, A. Buffa, G. Sangalli, and R. Vázquez. Mathematical analysis of variational isogeometric methods. *Acta Numerica*, 23:157–287, 2014.
2. J. A. Cottrell, T. J. R. Hughes, and Y. Bazilevs. *Isogeometric Analysis, Toward Integration of CAD and FEA*. Wiley, 2009.
3. R. A Dalrymple. A numerical model for periodic finite amplitude waves on a rotational fluid. *Journal of Computational Physics*, 24(1):29–42, 1977.
4. C. de Boor. *A practical guide to splines*. Springer-Verlag, New York, 2001.
5. M. C. Delfour and J. P. Zolésio. *Shapes and Geometries: Metrics, analysis, differential calculus, and optimization*. Society for Industrial and Applied Mathematics (SIAM), Philadelphia, PA, 2011.
6. M. Ehrnström, J. Escher, and E. Wahlén. Steady Water Waves with Multiple Critical Layers. *SIAM Journal on Mathematical Analysis*, 43(3):1436–1456, 2011.
7. G. Farin. *Curves and Surfaces for CAGD*. Academic Press, 1990.
8. H. Gomez and L. De Lorenzis. The variational collocation method. *Computer Methods in Applied Mechanics and Engineering*, 309:152–181, 2016.
9. M. D. Groves. Steady water waves. *Journal of Nonlinear Mathematical Physics*, 11(4):435–460, 2004.
10. T. J. R. Hughes, J. A. Cottrell, and Y. Bazilevs. Isogeometric analysis: CAD, finite elements, NURBS, exact geometry and mesh refinement. *Computer Methods in Applied Mechanics and Engineering*.
11. E. H. van Brummelen K. G. van der Zee and R. de Borst. Goal-oriented error estimation and adaptivity for free-boundary problems: the shape-linearization approach. *SIAM Journal on Scientific Computing*, 32(2):1093–1118, 2010.
12. K. T. Kärkkäinen and T. Tiihonen. Free surfaces: Shape sensitivity analysis and numerical methods. *International Journal for Numerical Methods in Engineering*, 44:1079–1098, 1999.
13. K. T. Kärkkäinen and T. Tiihonen. Shape calculus and free boundary problems. In *European Congress on Computational Methods in Applied Sciences and Engineering ECCOMAS*, 2004.
14. M. Montardini, G. Sangalli, and L. Tamellini. Optimal-order isogeometric collocation at Galerkin superconvergent points. *Computer Methods in Applied Mechanics and Engineering*, 316:741–757, 2017.
15. L. Piegl and W. Tiller. *The NURBS book*. Springer, 1997.
16. C. H. Rycroft and J. Wilkening. Computation of three-dimensional standing water waves. *Journal of Computational Physics*, 255:612–638, 2013.
17. D. Schillinger, J. A. Evans, A. Reali, M. A. Scott, and T. J. R. Hughes. Isogeometric Collocation: Cost Comparison with Galerkin Methods and Extension to Adaptive Hierarchical NURBS Discretizations. *Computer Methods in Applied Mechanics and Engineering*, 267:170–232, 2013.
18. J. A. Simmen and P. G. Saffman. Steady Deep-Water Waves on a Linear Shear Current. *Studies in Applied Mathematics*, 73(1):35–57, 1985.
19. J. Sokolowski and J. P. Zolesio. *Introduction to shape optimization: shape sensitivity analysis*. Springer series in computational mathematics, Springer, Berlin, 1992.
20. J. F. Toland. Stokes waves. *Topological Methods in Nonlinear Analysis*, 7(1):1–48, 1996.
21. K. G. van der Zee, G. J. van Zwieten, C. V. Verhoosel, and E. H. van Brummelen. Shape-Newton Method for Isogeometric Discretization of Free-Boundary Problems. In *MARINE 2011, IV International Conference on Computational Methods in Marine Engineering : selected papers : part III*, pages 85–102. Springer, 2013.

22. J. M. Vanden-Broeck. *Gravity-Capillary Free-Surface Flows*. Cambridge University Press, 2010.
23. R. Vázquez. A new design for the implementation of isogeometric analysis in Octave and Matlab: GeoPDEs 3.0. *Computers & Mathematics with Applications*, 72(3):523–554, 2016.
24. E. Wahlén. Steady water waves with a critical layer. *Journal of Differential Equations*, 246(6):2468–2483, 2009.

Approximately C^1-Smooth Isogeometric Functions on Two-Patch Domains

Agnes Seiler and Bert Jüttler

Abstract Motivated by the promising recent results concerning the construction of smooth isogeometric functions on multi-patch domains on bilinearly parameterized domains [14] or reparameterizations of more general domains [5], which, however, impose quite restrictive assumptions on the underlying domain, we propose two approaches to construct spaces $G_h^{1,\varepsilon}$ of approximately C^1-smooth isogeometric functions on general two-patch domains. The main idea is to work with C^0-continuous functions and to bound the jump of their gradients across the interface between neighboring patches. The constructions are based on two suitably chosen bilinear forms $\mathcal{B}_1$ and $\mathcal{B}_2$ and their eigenstructures, which lead to different bounds on the gradient jumps, respectively. We show that while the gradient jumps of the functions based on $\mathcal{B}_1$ fulfill a stricter bound, the functions themselves do not realize optimal convergence rates. Numerical experiments suggest that the functions based on $\mathcal{B}_2$ reach the optimal approximation order for solving second order problems. Furthermore, they are smooth enough to solve higher order problems such as the biharmonic equation. However, the bound on their gradient jump is mesh-size dependent.

1 Introduction

Isogeometric Analysis, introduced in 2005 by Hughes et al. [6], is an approach to numerical simulation via partial differential equations (PDEs). The computational domain is represented by a spline parameterization, which is called the geometry mapping. The discretization relies on isogeometric functions, which serve as test

---

A. Seiler
Doctoral Program Computational Mathematics, Johannes Kepler University Linz, Linz, Austria
e-mail: agnes.seiler@dk-compmath.jku.at

B. Jüttler (✉)
Institute of Applied Geometry, Johannes Kepler University Linz, Linz, Austria
e-mail: bert.juettler@jku.at

functions in the weak form of the problem. They are obtained by concatenating the basis functions that contribute to the parameterization of the geometry with the inverse geometry mapping. Hence, Isogeometric Analysis (IgA) does not need a triangulation of the domain. Since it directly uses a spline parameterization to define the discretization, it is said to close the gap between the CAD representation of the geometry and numerical analysis [6].

Another advantage of IgA consists in the increased smoothness of the discretization compared to standard finite elements. Within the patches, isogeometric functions are typically C^{p-1} smooth. This facilitates the discretization of higher order problems. While simple physical domains can be parameterized by a single geometry map, more complicated ones will be represented as a collection of several patches, each of which with its own parameterization. In this case, the multi-patch structure of the domain has to be taken into account, since the isogeometric functions are not automatically smooth across patch interfaces. Thus, appropriate coupling methods are required.

Standard coupling methods from the finite element literature carry over to multi-patch isogeometric discretizations. These methods work with broken Sobolev spaces, where weak differentiability across patch interfaces is not guaranteed. Suitable coupling terms are added to the weak form of a partial differential equation. For instance, the mortar method [2, 3], Nitsche Mortaring [16, 19] or the discontinuous Galerkin method [15, 17] perform the coupling via average and jump terms. Those terms need to be adapted to the order of the problem. While it suffices to consider the jump of function values for second order problems, fourth order problems require to take the difference of the normal derivatives into account. More generally, via the relation between a coercive bilinear form of an elliptic problem and its equivalent quadratic optimization problem, suitable methods from non-linear optimization can be applied to the coupling problem [8].

This paper explores a different approach, which is based on approximately smooth isogeometric test functions on the entire domain. Consequently, no modification of the variational form is required.

The coupling of isogeometric discretization across patch interfaces recently attracted substantial interest:

- C^0-coupling of isogeometric functions can be performed easily by identifying the coefficients of neighboring basis functions along an interface.
- The construction of C^1-smooth test functions, which are useful for higher order problems, is considerably more complicated. Recent results rely on the relation between geometric continuity of a graph surface and the smoothness of the associated functions [7, 14]. However, C^1-constructions are based on certain assumptions about the parameterization of the underlying domain, which are needed to ensure sufficient flexibility of the resulting discretizations. For instance, in [9, 14], the authors use bilinear or bilinear-like parameterizations. A reparameterization is needed for more general domains [5, 10]. A numerical approach to the computation of C^1-smooth discretization is presented in [4].

- These results have partially been extended to C^2-smooth isogeometric discretizations [11, 12, 13].

In order to avoid the limitation to bilinear-like parameterizations, we relax the construction by considering approximate C^1 smoothness of isogeometric functions on multi-patch domains. This enables us to generate function spaces on general (not bilinear-like) domains. Our construction is based on suitably chosen bilinear forms. More precisely, we explore two different forms and obtain two different function spaces. Starting from globally C^0-smooth functions, we provide bounds on the gradient jump of the corresponding approximately C^1-smooth isogeometric functions.

The remainder of this paper is organized as follows: Sect. 2 introduces the notation and the two different bilinear forms $\mathcal{B}_1$ and $\mathcal{B}_2$. The next section describes the construction of a space of approximately C^1-smooth isogeometric functions based on $\mathcal{B}_1$ and investigates its properties. In particular, we observe that the resulting space is not guaranteed to contain the trivially smooth functions. In order to address this deficiency, Sect. 4 describes another construction, which is based on the simplified bilinear form $\mathcal{B}_2$. Section 5 is devoted to numerical experiments concerning the approximation power and the dimension of the spaces. In particular, we will provide experiments suggesting that the functions we construct are smooth enough to solve fourth-order problems. Finally, we conclude the paper.

2 Preliminaries

We consider a planar two-patch domain $\Omega = \Omega^1 \cup \Omega^2 \subseteq \mathbb{R}^2$ with interface Γ between the individual patches Ω^1 and Ω^2, as depicted in Fig. 1. It is parameterized by a tensor-product B-spline mapping G via

$$G : \hat{\Omega} \to \Omega : (\xi_1, \xi_2) \mapsto \sum_{i \in \mathcal{I}} P_i \beta_i(\xi_1, \xi_2), \quad (\xi_1, \xi_2) \in \hat{\Omega} = [-1, 1] \times [0, 1], \tag{1}$$

where $P_i \in \mathbb{R}^2$ are control points and β_i are tensor-product B-splines of bidegree (p_1, p_2) with index set $\mathcal{I}$, defined by open knot vectors Ξ_1, Ξ_2 with maximal knot span sizes h_1, h_2 in ξ_1 and ξ_2 direction, respectively. We set $h = \max\{h_1, h_2\}$. The multiplicities of the inner knots do not exceed $p - 1$, except for the knot 0 in Ξ_1, which appears p times. The simplest instance of the knot configuration is visualized in Fig. 1. The patch interface is $\Gamma = G(\{0\} \times [0, 1])$. The associated isogeometric basis functions

$$b_i(\mathbf{x}) = \left(\beta_i \circ G^{-1}\right)(\mathbf{x}), \; i \in \mathcal{I} \tag{2}$$

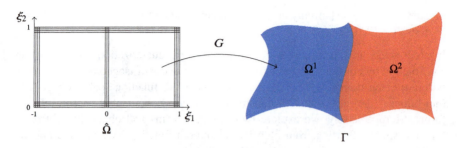

Fig. 1 Two patch domain Ω parameterized by a bicubic geometry map G. The knot vectors are given by $[-1, -1, -1, -1, 0, 0, 0, 1, 1, 1, 1] \times [0, 0, 0, 0, 1, 1, 1, 1]$

are collected in the vector

$$b(x) = \bigl(b_i(x)\bigr)_{i \in \mathcal{I}} \tag{3}$$

and span the isogeometric discretization space

$$\mathcal{V}_h = \text{span}\{b_i : i \in \mathcal{I}\} \subseteq C^0(\Omega). \tag{4}$$

Finally we recall the definition of the jump operator

$$[f] = f^1|_\Gamma - f^2|_\Gamma,$$

which is defined for any function $f \in L^2(\Omega)$ with

$$f^1 = f|_{\Omega^1} \in H^1(\Omega^1), \quad f^2 = f|_{\Omega^2} \in H^1(\Omega^2).$$

We will use two different bilinear forms in order to construct approximately C^1-smooth isogeometric functions on Ω. The first one is given by

$$\mathcal{B}_1 : \mathcal{V}_h \times \mathcal{V}_h \to \mathbb{R} : (f, g) \mapsto \varepsilon \int_\Omega f(x)g(x)dx - \int_\Gamma [\nabla f(x)]^T [\nabla g(x)]dx \tag{5}$$

and depends on a positive parameter ε. The second one takes the form

$$\mathcal{B}_2 : \mathcal{V}_h \times \mathcal{V}_h \to \mathbb{R} : (f, g) \mapsto \int_\Gamma [\nabla f(x)]^T [\nabla g(x)]dx. \tag{6}$$

More precisely, our aim is to construct isogeometric functions with a bounded gradient jump $\|[\nabla f]\|_{L^2(\Gamma)}$, and the bilinear forms $\mathcal{B}_1$ and $\mathcal{B}_2$ are designed with this objective in mind. For the first one, a suitable value of ε has to be chosen in advance. It controls the magnitude of the bound. The spaces of approximately smooth isogeometric functions obtained by using $\mathcal{B}_1$ and $\mathcal{B}_2$ have different properties, although the constructions themselves are quite similar.

3 Results for $\mathcal{B}_1$

We show that the bilinear form $\mathcal{B}_1$ yields a mesh-size independent bound on the gradient jump. However, we will also see that this space lacks optimal approximation power.

3.1 Construction of Approximately Smooth Functions

We consider functions $f, g \in \mathcal{V}_h$ with

$$f(x) = u^T b(x), g(x) = v^T b(x), \tag{7}$$

with coefficient vectors $u, v \in \mathbb{R}^{|\mathcal{I}|}$. Consequently, $\mathcal{B}_1(f, g)$ can be rewritten in matrix-vector-form as

$$\mathcal{B}_1(f, g) = u^T (\varepsilon M - Q) v, \tag{8}$$

where

$$M = (m_{i,j})_{i,j \in \mathcal{I}} \text{ with } m_{i,j} = \int_\Omega b_i(x) b_j(x) dx \tag{9}$$

and

$$Q = (q_{i,j})_{i,j \in \mathcal{I}} \text{ with } q_{i,j} = \int_\Gamma [\nabla b_i(x)]^T [\nabla b_j(x)] dx, \tag{10}$$

as confirmed by a short computation. The matrices M and Q are symmetric positive semi-definite, as

$$u^T M u = \|f\|_{L^2(\Omega)}^2 \geq 0 \tag{11}$$

and

$$u^T Q u = \|[\nabla f]\|_{L^2(\Gamma)}^2 \geq 0. \tag{12}$$

Now let $0 \leq \lambda^1 \leq \ldots \leq \lambda^n$ be the non-negative eigenvalues of $\varepsilon M - Q$ in ascending order and let $c^1, \ldots, c^n$ be the corresponding eigenvectors, $n \leq |\mathcal{I}|$. The eigenvectors satisfy

$$(c^k)^T c^\ell = 0 \text{ and } (c^k)^T c^k = 1 \text{ for } k \neq \ell, 1 \leq k, \ell \leq n, \tag{13}$$

possibly after performing the Gram-Schmidt orthonormalization, if multiple eigenvalues are present.

We define

$$\begin{aligned} G_h^{1,\varepsilon} &:= \text{span}\left\{ \sum_{i \in I} c_i^k b_i(\mathbf{x}) : k = 1, \ldots, n \right\} \\ &= \left\{ \sum_{i \in I} d_i b_i(\mathbf{x}) : d \in \text{span}\left\{ c^1, \ldots c^n \right\} \right\}. \end{aligned} \tag{14}$$

as the *space of approximately C^1-smooth isogeometric functions*.

3.2 Properties of the Function Space

By construction, $G_h^{1,\varepsilon}$ is a linear space. As an immediate consequence from its definition, we obtain a mesh-size independent bound on the gradient jump.

We denote by $\mathbf{C} \in \mathbb{R}^{|I| \times n}$ the matrix containing the eigenvectors $c^1, \ldots, c^n$ as column vectors.

Proposition 1 *The gradient jump of any function $f = (\mathbf{C}d)^T \mathbf{b} \in G_h^{1,\varepsilon}$ with $d \in \mathbb{R}^n$ can be bounded by*

$$\|[\nabla f]\|_{L^2(\Gamma)}^2 \leq \varepsilon \|f\|_{L^2(\Omega)}^2. \tag{15}$$

Proof We use (11) and (12) and obtain

$$\begin{aligned} \varepsilon \|f\|_{L^2(\Omega)}^2 - \|[\nabla f]\|_{L^2(\Gamma)}^2 &= (\mathbf{C}d)^T (\varepsilon M - Q)(\mathbf{C}d) \\ &= d^T \mathbf{C}^T (\varepsilon M - Q) \mathbf{C} d \\ &= d^T \text{diag}\left(\lambda^1, \ldots, \lambda^n\right) d \\ &= \lambda^1 d_1^2 + \ldots + \lambda^n d_n^2 \geq 0, \end{aligned}$$

where the last inequality holds because we only consider non-negative eigenvalues λ_i. □

The space $G_h^{1,\varepsilon}$ based on $\mathcal{B}_1$ does not necessarily contain the trivially smooth isogeometric functions b_i^k with

$$\nabla b_i^k|_\Gamma = \mathbf{0}.$$

We will refer to these functions as *off-interface basis functions*. Since they are constantly zero across the interface Γ, their gradient jumps across Γ are zero as well. However, their coefficient vectors with respect to the basis $\boldsymbol{b}$, which are the canonical unit vectors in $\mathbb{R}^{|\mathcal{I}|}$, are not necessarily eigenvectors of $\varepsilon M - Q$. As we shall see in Sect. 5, the functions in $G_h^{1,\varepsilon}$ do not possess the same approximation power as the full space of isogeometric functions.

4 Results for $\mathcal{B}_2$

We study another bilinear form in order to ensure the existence of trivially smooth functions in the resulting space of approximately smooth isogeometric functions. However, in this case we cannot expect to obtain an estimate of $\|[\nabla f]\|_{L^2(\Gamma)}$ that is independent of the mesh size.

4.1 Construction of Approximately Smooth Functions

The modified space $\hat{G}_h^{1,\varepsilon}$ is constructed analogously to the procedure described in Sect. 3.1. Recall that $\mathcal{B}_2(f, f)$ can equivalently be written as

$$\mathcal{B}_2(f, f) = u^T Q u$$

for

$$f = u^T \boldsymbol{b} \in \mathcal{V}_h, \quad u \in \mathbb{R}^{|\mathcal{I}|}.$$

As explained before, the matrix Q is symmetric positive semi-definite. We choose a positive value ε. Let $\hat{\lambda}^1 \leq \ldots \leq \hat{\lambda}^{\hat{n}} \leq \varepsilon$ be the eigenvalues of Q that are bounded by ε and let $\hat{c}^1, \ldots, \hat{c}^{\hat{n}}$ be the corresponding orthonormalized eigenvectors, $\hat{n} \leq |\mathcal{I}|$. We define

$$\hat{G}_h^{1,\varepsilon} := \mathrm{span}\left\{\sum_{i \in \mathcal{I}} \hat{c}_i^k b_i : k = 1, \ldots, \hat{n}\right\}. \tag{16}$$

4.2 Properties of the Function Space

Again, by construction, $\hat{G}_h^{1,\varepsilon}$ is a linear space. Moreover, all trivially smooth isogeometric functions f, i.e., the off-interface basis functions as well as constant and linear functions (which are contained in the space of isogeometric functions, due to use of the isoparametric principle), fulfill

$$\mathcal{B}_2(f, f) = 0.$$

Since the matrix Q is symmetric positive semi-definite, this implies that the coefficient vector of f is an element of the kernel of Q. Consequently, the corresponding coefficient vector is an eigenvector to the eigenvalue 0 of Q. Since we set $\varepsilon > 0$, all elements in the kernel will also be elements of $\hat{G}_h^{1,\varepsilon}$. This is independent of the mesh size h. As we will see, the inclusion of these functions in $\hat{G}_h^{1,\varepsilon}$ is important to achieve optimal convergence.

Subsequently, we bound the gradient jump of functions in $\hat{G}_h^{1,\varepsilon}$. We denote by $\hat{C} \in \mathbb{R}^{|I| \times \hat{n}}$ the matrix containing the eigenvectors $\hat{c}^1, \ldots, \hat{c}^{\hat{n}}$ of Q as column vectors. Let $f \in \hat{G}_h^{1,\varepsilon}$, i.e. we set

$$f(x) = (\hat{C}d)^T b(x) \tag{17}$$

with $d \in \mathbb{R}^n$.

Theorem 1 *Let the knot vectors Ξ_1, Ξ_2 be quasi-uniform. Then all functions $f \in \hat{G}_h^{1,\varepsilon}$ satisfy*

$$\|[\nabla f]\|_{L^2(\Gamma)}^2 \leq \varepsilon \frac{C}{h^2} \|f\|_{L^2(\Omega)}^2 \tag{18}$$

for a constant C that depends on the maximal spline degree p and the geometry mapping G, but not on the maximal mesh size h.

Proof Let $f = (\hat{C}d)^T b \in \hat{G}_h^{1,\varepsilon}$ as denoted above. Then we have

$$\|[\nabla f]\|_{L^2(\Gamma)}^2 = (\hat{C}d)^T Q(\hat{C}d) = d^T \hat{C}^T Q \hat{C} d$$

$$= d^T \operatorname{diag}\left(\hat{\lambda}^1, \ldots, \hat{\lambda}^{\hat{n}}\right) d$$

$$= \hat{\lambda}^1 d_1^2 + \ldots + \hat{\lambda}^{\hat{n}} d_{\hat{n}}^2$$

$$\leq \varepsilon \sum_{i=1}^{\hat{n}} d_i^2 = \varepsilon \|d\|_2^2 = \varepsilon \|\hat{C}d\|_2^2,$$

where the last equality holds because $\hat{C}$ is an orthogonal matrix.

Next, we use the stability of tensor-product B-spline bases $\{\beta_i\}_{i\in I}$ [18, Theorem 12.5], which describes a relation between the coefficients of a spline function and the function itself, in the form

$$\|\hat{C}d\|_2 \leq D_p^2 \frac{1}{h} \left\| \sum_{i\in I} (\hat{C}d)_i \beta_i \right\|_{L^2(\hat{\Omega})} \tag{19}$$

with stability constant D_p^2, where $p = \max\{p_1, p_2\}$. Hence, we get

$$\|[\nabla f]\|_{L^2(\Gamma)}^2 \leq \varepsilon \|\hat{C}d\|_2^2 \leq \varepsilon D_p^4 \frac{1}{h^2} \left\| \sum_{i\in I} (\hat{C}d)_i \beta_i \right\|_{L^2(\hat{\Omega})}^2 . \tag{20}$$

We rewrite β_i in terms of the push-forward $b_i \circ G$ and obtain

$$\|[\nabla f]\|_{L^2(\Gamma)}^2 \leq \varepsilon D_p^4 \frac{1}{h^2} \left\| \sum_{i\in I} (\hat{C}d)_i (b_i \circ G) \right\|_{L^2(\hat{\Omega})}^2 , \tag{21}$$

which again can be rewritten and summarized as

$$\|[\nabla f]\|_{L^2(\Gamma)}^2 \leq \varepsilon D_p^4 \frac{1}{h^2} \left\| \left(\sum_{i\in I} (\hat{C}d)_i b_i \right) \circ G \right\|_{L^2(\hat{\Omega})}^2 \tag{22}$$

$$= \varepsilon D_p^4 \frac{1}{h^2} \|f \circ G\|_{L^2(\hat{\Omega})}^2 .$$

Now we transform the integral $\int_{\hat{\Omega}} (f \circ G)^2$ on $\hat{\Omega}$ to an integral on Ω, which yields

$$\|[\nabla f]\|_{L^2(\Gamma)}^2 \leq \varepsilon D_p^4 \frac{1}{h^2} \|\det \nabla (G)^{-1}\|_{L^\infty(\Omega)} \|f\|_{L^2(\Omega)}^2. \tag{23}$$

Finally we set

$$C(p, G) = D_p^4 \cdot \left\| \det \nabla(G)^{-1} \right\|_{L^\infty(\Omega)}.$$

This concludes the proof. □

This result resembles standard inverse inequalities for isogeometric functions, which can be found in [1], apart from the power of h and the factor ε, which is chosen in advance. If we chose $\varepsilon \in O(h^2)$, we can eliminate the mesh-size dependence in the bound of the gradient jump. However, smaller values of ε lead to fewer functions in $\hat{G}_h^{1,\varepsilon}$, which we will discuss in the following section.

5 Numerical Examples

We consider least squares approximation, the Poisson problem, and the biharmonic equation on a two-patch domain. In this context we are interested in the approximation power of $G_h^{1,\varepsilon}$ and $\hat{G}_h^{1,\varepsilon}$. Furthermore we will study the number of interface basis functions under uniform h-refinement.

5.1 Approximation Power

Throughout the remainder of this section, all errors are measured patch-wisely and then summed up, e.g. we refer to

$$\| f_{\text{approx}}|_{\Omega^1} - f_{\text{exact}}|_{\Omega^1} \|_{H^1(\Omega^1)} + \| f_{\text{approx}}|_{\Omega^2} - f_{\text{exact}}|_{\Omega^2} \|_{H^1(\Omega^2)}$$

as the H^1 error and to

$$\| f_{\text{approx}}|_{\Omega^1} - f_{\text{exact}}|_{\Omega^1} \|_{H^2(\Omega^1)} + \| f_{\text{approx}}|_{\Omega^2} - f_{\text{exact}}|_{\Omega^2} \|_{H^2(\Omega^2)}$$

as the H^2 error of f_{approx}. The patch-wise splitting is not necessary for the L^2 error, as $G_h^{1,\varepsilon} \subseteq L^2(\Omega)$ and $\hat{G}_h^{1,\varepsilon} \subseteq L^2(\Omega)$.

5.1.1 Least Squares Approximation

We start with an example that identifies the limitations of the space $G_h^{1,\varepsilon}$, which is based on the bilinear form $\mathcal{B}_1$. Figure 2 shows the function

$$f_{\text{exact}}(x, y) = 3xy \exp(-x) \sin(\pi y) \tag{24}$$

which we approximate on a two-patch domain by functions in $G_h^{1,\varepsilon}$. The domain coincides with the one shown in Fig. 1. We solve the constrained least squares fitting problem

$$\min_{f \in G_h^{1,\varepsilon}} \| f - f_{\text{exact}} \|_{L^2(\Omega)}^2.$$

Fig. 2 Bicubically parameterized domain (see Fig. 1) and transparent plot of the exact solution $3xy \exp(-x) \sin(\pi y)$

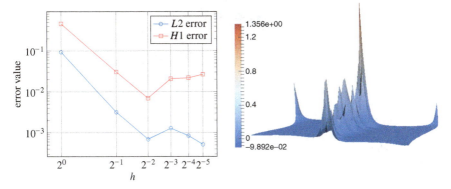

Fig. 3 Least squares approximation with functions in $G_h^{1,\varepsilon}$ with $\varepsilon = 0.5$. Left: relative L^2 and H^1 errors for the approximate solution of degree 3. Right: L^2 error values at the finest discretization step (scaled by factor 100)

The parameter ε was set to 0.5. The relative L^2 and H^1 errors are depicted in the left plot of Fig. 3. After some refinement steps, no significant reduction of the error is achieved. Considering the distribution of the error values in the last refinement step, shown in Fig. 3, right, we note that the largest errors occur close to the interface and in the back corners. This is a possible indicator that the corresponding corner basis functions are not present in $G_h^{1,\varepsilon}$.

Consequently, we consider only the space $\hat{G}_h^{1,\varepsilon}$ based on the bilinear form $\mathcal{B}_2$. The following experiment shows that - in contrast to the previous approach - the functions in $\hat{G}_h^{1,\varepsilon}$ maintain the full approximation power.

Again, we choose $\varepsilon = 0.5$ and approximate the same function (24) on the same domain as before. We use a uniform h-refinement strategy. The relative L^2 and H^1 error values and the respective convergence rates are shown in Fig. 4, top left and top right. A comparison with the reference slopes shows that the functions in $\hat{G}_h^{1,\varepsilon}$ maintain the optimal convergence rates of $p+1$ and p for the L^2 and the H^1 error, respectively.

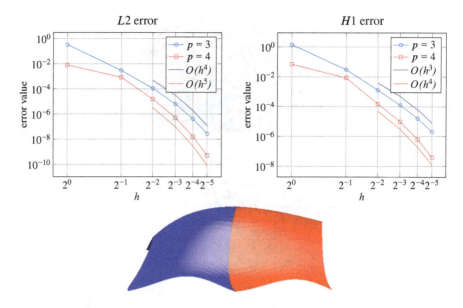

Fig. 4 Least squares approximation with functions in $\hat{G}_h^{1,\varepsilon}$ with $\varepsilon = 0.5$. Relative L^2 (top left) and H^1 (top right) errors of the solution to the fitting problem on the bi-cubic domain, see Fig. 1. Bottom: patch-wise representation of the solution with 8840 basis functions of degree three with flat shading effects

At the finest level of refinement we used 8840 (9111) basis functions of degree 3 (4) with mesh size 2^{-6}. Note that this number of basis functions is slightly less than the number of original tensor-product B-splines, which is 8978 (9248) for degree 3 (4).

The bottom plot in Fig. 4 depicts the solution using 8840 basis functions of degree 3 as a patch-wise plot with added flat shading effects. These effects highlight the smoothness of the solution across the curved interface.

5.1.2 Poisson Problem

Solving the Poisson equation leads to very similar results. The physical domain Ω, on which we study the problem, consists of two patches with a curved interface, see Fig. 5, left. It is bi-quadratically parameterized. We consider the discretized weak form:

$$\text{Find } u \in \hat{G}_{h,0}^{1,\varepsilon} \text{ such that } \int_\Omega \nabla u(\boldsymbol{x}) \nabla v(\boldsymbol{x}) d\boldsymbol{x} = \int_\Omega f(\boldsymbol{x}) v(\boldsymbol{x}) d\boldsymbol{x} \quad \forall v \in \hat{G}_{h,0}^{1,\varepsilon}, \tag{25}$$

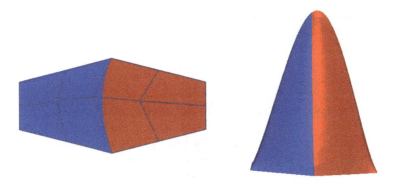

Fig. 5 Poisson problem (25). Left: Domain Ω and its control net. Right: Patch-wise plot of the solution with 2048 basis functions of degree two in $\hat{G}_h^{1,\varepsilon}$ for $\varepsilon = 0.5$ with flat shading effects

where $\hat{G}_{h,0}^{1,\varepsilon} = \{u \in \hat{G}_h^{1,\varepsilon} : u|_{\partial\Omega} = 0\}$, i.e., the zero Dirichlet boundary conditions are imposed strongly in the test function space. Again we set the threshold ε to 0.5. The right hand-side f was chosen as the negative Laplacian of the exact solution is given by

$$u(x, y) = 40(0.25x + 0.75 - y)(-0.25x + 1.25 - y)$$
$$(-0.25x + 0.25 - y)(0.25x - 0.25 - y)\sin(0.5\pi x).$$

The solution to (25) is found by means of a Galerkin method. Figure 5, right, shows its solution for 2048 degrees of freedom with element size 2^{-5}. The patch-wise plot with the flat shading effect emphasizes that the solution is smooth in the area of the interface.

The behavior of the relative L^2 and H^1 errors is shown in Fig. 6, left and right, respectively. We see that in both cases and for the tested degrees two, three and four

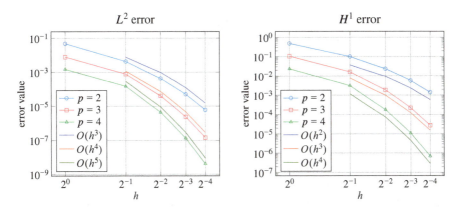

Fig. 6 Poisson problem (25). Relative L^2 (left) and H^1 (right) errors of the approximate solution for basis functions of different degrees in $\hat{G}_h^{1,\varepsilon}$ for $\varepsilon = 0.5$

of test functions we realize optimal convergence rates. This is consistent with the L^2 approximation results.

5.1.3 Biharmonic Equation

The examples shown previously did not require C^1-smooth basis functions. Consequently, the approximately smooth functions we presented did not exhibit any advantage over standard C^0-smooth isogeometric functions (which can be constructed by identifying the corresponding degrees of freedom along the interface), except for the fact that we used slightly less basis functions. We now consider a fourth-order equation, where the bilinear form governing the weak formulation cannot be evaluated for only C^0-smooth functions. The following examples demonstrate that—depending on the value of the parameter ε that controls the magnitude of the jump—approximately smooth functions are suitable for solving such a higher-order problem and even to maintain full approximation power with respect to the L^2, H^1 and H^2 errors.

We consider the discretized weak form of the biharmonic equation:

$$\text{Find } u \in \hat{G}_{h,0}^{1,\varepsilon} \text{ such that } \int_\Omega \Delta u(x) \Delta v(x) d(x) = \int_\Omega f(x) v(x) \quad \forall v \in \hat{G}_{h,0}^{1,\varepsilon}, \tag{26}$$

where $\hat{G}_{h,0}^{1,\varepsilon} = \{u \in \hat{G}_h^{1,\varepsilon} : u|_{\partial\Omega} = (\nabla u \cdot n)|_{\partial\Omega} = 0\}$. Again, we impose the boundary conditions strongly in the test function space and solve (26) by means of the Galerkin method. The right-hand side f is obtained from the exact solution $(1 - \cos(2\pi x))(1 - \cos(2\pi y))$. The domain Ω is a square, which is split into two patches with a curved interface, see Fig. 7, left. Figure 7, right, depicts the solution for 2101 basis functions of degree four and element size $h = 2^{-5}$ for $\varepsilon = h^2$. The shading demonstrates the smoothness of the solution across the interface in the patch-wise plot.

We consider the decay of the relative error for different degrees of the basis functions, starting with degree $p = 3$. The plots in Fig. 8 show that the optimal approximation order with respect to the L^2 (left) and H^2 (right) norm is reached for $\varepsilon = C \cdot h^k$ for $k \leq 2$, but not for $k = 3$.

The situation is slightly different for $p = 4$. Here, the optimal approximation order with respect to the L^2 (left) and H^2 (right) norm is reached for $\varepsilon = C \cdot h^k$ for $k = 2, 3$, but neither for $k \leq 1$ nor for $k \geq 4$, see Fig. 9. Finally, the optimal approximation order for $p = 5$ with respect to the L^2 (left) and H^2 (right) norm is reached for $\varepsilon = C \cdot h^k$ for $k = 3$, but neither for $k \leq 2$ nor for $k \geq 4$, as shown in Fig. 10.

On the one hand, a higher power of h and thus a smaller value of ε results in smoother, but at the same time in fewer basis functions, hence in a loss of approximation power. On the other hand, while choosing a larger value of ε

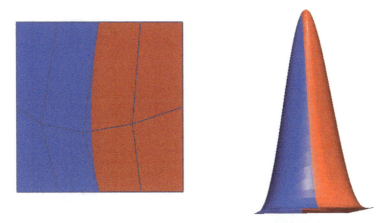

Fig. 7 Biharmonic equation (26): Domain with its control net (left) and patch-wise plot of the solution with shading (right) for 2101 basis functions of degree four in $\hat{G}_h^{1,\varepsilon}$ with $\varepsilon = h^2$

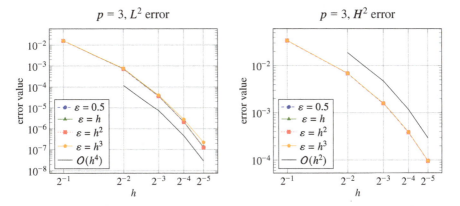

Fig. 8 Biharmonic equation (26): Relative L^2 (left) and H^2 (right) errors of the approximate solution for basis functions of degree three in $\hat{G}_h^{1,\varepsilon}$ for four choices of ε

increases the dimension of $\hat{G}_h^{1,\varepsilon}$, the resulting discretizations are not smooth enough for solving higher order problems. We conjecture that $\varepsilon = C \cdot h^{p-2}$ is the optimal choice.

5.2 Dimension of the Space

We investigate the influence of ε on the number of interface basis functions, and thus on the dimension of the space $\hat{G}_h^{1,\varepsilon}$. Note that the number of trivially smooth basis functions is not affected by the choice of ε.

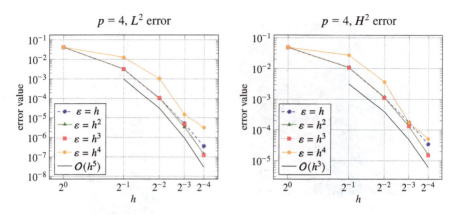

Fig. 9 Biharmonic equation (26): Relative L^2 (left) and H^2 (right) errors of the approximate solution for basis functions of degree four in $\hat{G}_h^{1,\varepsilon}$ for four choices of ε

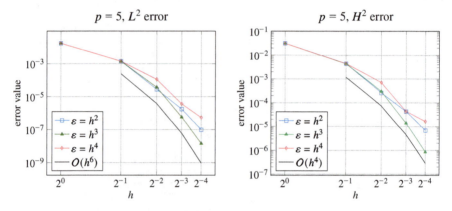

Fig. 10 Biharmonic equation (26): Relative L^2 (left) and H^2 (right) errors of the numeric solution for basis functions of degree five in $\hat{G}_h^{1,\varepsilon}$ for three choices of ε

We cannot expect nested spaces, i.e., we cannot ensure that

$$\hat{G}_h^{1,\varepsilon} \subseteq \hat{G}_{\frac{h}{2}}^{1,\varepsilon}.$$

Nevertheless, the number of interface basis functions grows as h is decreased.

Figure 11 shows the number of interface basis functions for different degrees and different choices of ε. For $\varepsilon = C \cdot h^k$ with $k \leq p - 2$, the number of interface basis functions grows linearly under h-refinement for all degrees. A larger choice of k, however, results in significantly fewer functions.

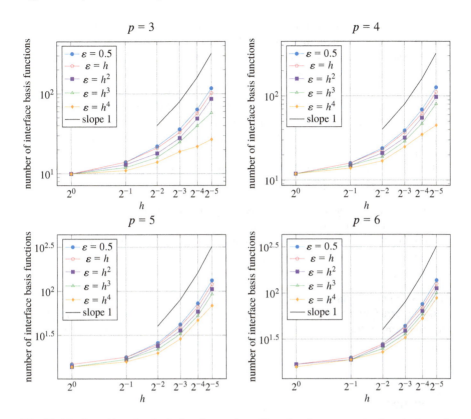

Fig. 11 Number of interface basis functions under uniform h-refinement on the domain shown in Fig. 1. Basis functions of degree three (top left), four (top right), five (bottom left) and six (bottom right) for various choices of ε

6 Conclusion

We proposed a concept of constructing approximately C^1-smooth isogeometric functions on planar multi-patch domains which is based on selecting eigenvalues and corresponding eigenvectors of the matrix representation of a suitable bilinear form. The functions will have a non-zero gradient jump across the interface.

We studied two different bilinear forms. The first bilinear form led to a bound on the gradient jump of the form

$$\|[\nabla f]\|_{L^2(\Gamma)}^2 \leq \varepsilon \|f\|_{L^2(\Omega)}^2,$$

where ε is to be chosen in advance. This bound is h-independent. However, the space constructed via this bilinear form does not necessarily contain trivially smooth functions, which led to a decrease in the approximation order. The function space based on the second bilinear form contains all trivially smooth isogeometric functions and the gradient jump is bounded by a mesh-dependent term.

Numerical experiments suggested that for second-order problems, the latter approach maintains the optimal approximation order even for constant choices of ε and that the functions are sufficiently smooth to solve the biharmonic equation. The convergence of the approximate solution was influenced by the choice of ε. Depending on the degree of the basis functions, ε had to be chosen as a suitable power of the mesh size h in order to achieve convergence of the solution to the biharmonic problem.

In future work we would like to establish a theoretical background for the experimental results. This includes

- investigating a projector to the space of approximately C^1-smooth isogeometric functions to prove optimal convergence rates,
- studying the eigenstructure of the matrix Q to develop a lower bound for the number of non-trivial basis functions and
- analyzing the influence of ε.

The last point affects the first two points as well: A smaller value of ε creates smoother but fewer functions. Last but not least we are interested in generalizing the approach to domains with more than two patches and to the three-dimensional case.

References

1. Y. Bazilevs, L. B. de Veiga, J. A. Cottrell, T. J. Hughes, and G. Sangalli. Isogeometric analysis: Approximation, stability and error estimates for h-refined meshes. *Mathematical Models and Methods in Applied Sciences*, 16:1031–1090, 2006.
2. C. Bernardi, Y. Maday, and F. Rapetti. Basics and some applications of the mortar element method. *GAMM-Mitteilungen*, 28(2):97–123, 2005.
3. E. Brivadis, A. Buffa, B. Wohlmuth, and L. Wunderlich. Isogeometric mortar methods. *Computer Methods in Applied Mechanics and Engineering*, 284:292–319, 2015.
4. C. L. Chan, C. Anitescu, and T. Rabczuk. Isogeometric analysis with strong multipatch C^1-coupling. *Computer Aided Geometric Design*, 62:294–310, 2018.
5. A. Collin, G. Sangalli, and T. Takacs. Analysis-suitable G^1 multi-patch parametrizations for C^1 isogeometric spaces. *Computer Aided Geometric Design*, 47:93–113, 2016.
6. J. A. Cottrell, T. J. R. Hughes, and Y. Bazilevs. *Isogeometric Analysis. Toward Integration of CAD and FEA*. John Wiley and Sons, Chichester, England, 2009.
7. D. Groisser and J. Peters. Matched G^k-constructions always yield C^k-continuous isogeometric elements. *Computer Aided Design*, 34:67–72, 2015.
8. C. Grossmann. Penalties, Lagrange multipliers and Nitsche mortaring. *Differential Inclusions, Control and Optimization*, 20:205–220, 2010.
9. M. Kapl, F. Buchegger, M. Bercovier, and B. Jüttler. Isogeometric analysis with geometrically continuous functions on planar multi-patch geometries. *Computer Methods in Applied Mechanics and Engineering*, 316:209–234, 2017.
10. M. Kapl, G. Sangalli, and T. Takacs. Dimension and basis construction for analysis-suitable G^1 two-patch parameterizations. *Computer Aided Geometric Design*, 52:75–89, 2017.

11. M. Kapl and V. Vitrih. Space of C^2-smooth geometrically continuous isogeometric functions on planar multi-patch geometries: Dimension and numerical experiments. *Computers & Mathematics with Applications*, 73:2319–2338, 2017.
12. M. Kapl and V. Vitrih. Space of C^2-smooth geometrically continuous isogeometric functions on two-patch geometries. *Computers & Mathematics with Applications*, 73(1):37–59, 2017.
13. M. Kapl and V. Vitrih. Dimension and basis construction for C^2-smooth isogeometric spline spaces over bilinear-like G^2 two-patch parameterizations. *Journal of Computational and Applied Mathematics*, 335:289 – 311, 2018.
14. M. Kapl, V. Vitrih, B. Jüttler, and K. Birner. Isogeometric analysis with geometrically continuous functions on two-patch geometries. *Computers & Mathematics with Applications*, 70(7):1518–1538, 2015.
15. U. Langer and I. Toulopoulos. Analysis of multipatch discontinuous Galerkin IgA approximations to elliptic boundary value problems. *Computing and Visualization in Science*, 17(5):217–233, 2016.
16. V. P. Nguyen, P. Kerfriden, M. Brino, S. P. Bordas, and E. Bonisoli. Nitsche's method for two and three dimensional NURBS patch coupling. *Computational Mechanics*, 53(6):1163–1182, 2014.
17. B. Rivière. *Discontinuous Galerkin Methods for Solving Elliptic and Parabolic Equations: Theory and Implementation*. SIAM, 2008.
18. L. L. Schumaker. *Spline Functions: Basic Theory*. Wiley, New York, 1981.
19. M. F. Wheeler. An elliptic collocation-finite element method with interior penalties. *SIAM Journal on Numerical Analysis*, 15(1):152–161, 1978.

Properties of Spline Spaces Over Structured Hierarchical Box Partitions

Ivar Stangeby and Tor Dokken

Abstract Given a spline space spanned by Truncated Hierarchical B-splines (THB), it is always possible to construct a spline space spanned by Locally Refined B-splines (LRB) that contains the THB-space. Starting from configurations where the two spline spaces are equal, we address what happens to the properties of the LRB-space when it is modified by local one-directional refinement at convex corners of, and along edges between dyadic refinement regions. We show that such local modifications can reduce the number of B-splines over each element to the minimum prescribed by the polynomial bi-degree, and that such local refinements can be used for improving the condition numbers of mass and stiffness matrices.

1 Introduction

The use of Hierarchical B-splines (HB) introduced in [2] has gained much attention in Isogeometric Analysis (IgA) in recent years. Hierarchical B-splines are based on a dyadic sequence of grids determined by scaled lattices. On each hierarchical level a spline space is defined as the tensor product of univariate spline spaces spanned by uniform B-splines.

Hierarchical B-splines do not constitute a partition of unity, a much desired property in both Computer Aided Design (CAD) and IgA. As a remedy to this Truncated Hierarchical B-splines (THB) [5, 13] were introduced, where B-splines on one hierarchical level are suitably *truncated* by B-splines from finer hierarchical levels when the support of a B-spline at a finer level is contained in the support of a B-spline at a coarser level.

An alternative to the THB-approach for forming a partition of unity came with the introduction of Locally Refined B-splines (LRB) [1], where initial tensor product B-splines are split until only B-splines of minimal support remain. LRB permits

---

I. Stangeby · T. Dokken (✉)
SINTEF, Oslo, Norway
e-mail: ivar.stangeby@sintef.no; tor.dokken@sintef.no

dyadic refinement of hierarchical meshes while ensuring that all B-splines have minimal support. In the occasional case where meshlines at a dyadic level are too short to split an LR B-spline, the meshlines in question are extended. This fact ensures that the spline space spanned by THB-splines is either identical to or constitutes a subset of the LRB spline space.

In IgA *open knot vectors* are used to simplify the interpolation of boundary conditions, as reported for THB in [4] and for LRB in [6]. In *open knot vectors* the multiplicity at boundary knots is set to $m = d + 1$. An alternative approach is to use B-splines with knot multiplicity of $m = 1$ along the boundary. In order to force the partition of unity in this case, a ghost domain is added around the domain of interest, as seen in [7] for both THB and LRB. This distinction is illustrated for univariate cubic splines in Fig. 1.

In Sect. 2 we address the effects the choice of boundary knot multiplicity has on condition numbers. To distinguish between single multiplicity and open knots at the domain boundary we prefix any method using single knot multiplicity on the boundary with a ghosted domain by an "S". Using this naming convention, the methods addressed in [6] are respectively S-THB and S-LRB. In this paper, we show that for the same tensor product spline space, THB and LRB are superior with respect to condition numbers of mass and stiffness matrices compared to respectively S-LRB and S-THB. We also explain the intriguing near constant behaviour of the condition numbers reported in [7], where S-LRB and S-THB were addressed, and condition numbers seemed to be nearly independent of the refinement level. We show that this is due to single knot multiplicity at domain boundaries for the examples presented in [7]. Further it is shown that for more levels of refinement the condition numbers for the mass matrix for S-THB and S-LRB will meet and then follow the growing curves for respectively THB and LRB.

In HB and THB the refinement procedure (at an element level) consists of marking elements for splitting. Marked elements are subsequently split in both parameter directions. This contrasts with the refinement procedure LRB allows, namely that of splitting an element in a single parameter direction at the time, provided that at least one B-spline is split in the process. This can be used to modify the hierarchical refinement, and possibly improve the approximation properties of the resulting spline space. In the remaining sections we use open knot vectors at domain boundaries and address how such modifications influence the condition numbers for mass and stiffness matrices for bi-cubic spline spaces in particular. The remaining sections are structured as follows:

Section 3 gives a lightweight introduction to box-partitions and spline spaces over such partitions. The starting point for THB and LRB refinement is a tensor product spline space. The key concept of element overloading is defined, the situation where more B-splines cover an element than are needed for spanning the polynomial space over the element. We briefly summarize some key properties. Subsequently, we recall the definitions of both LRB and THB splines. We also relate the refinement strategies for LRB to T-splines [12]. Those readers already familiar with the contents of this section may feel free to skip it.

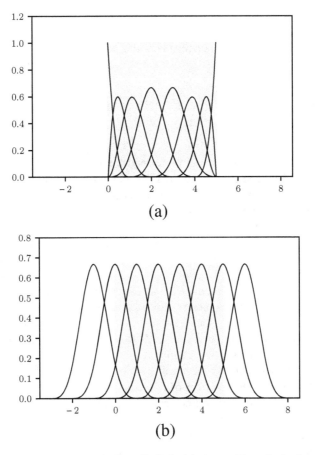

Fig. 1 Spline spaces over the domain $\Omega = [0, 5]$. In (**a**), the partition of unity is satisfied at the boundary by setting the knot multiplicity to $m = d+1 = 4$. In (**b**), the partition of unity is satisfied at the boundary by extending the domain to allow the full polynomial space to be spanned at the boundary elements. The shaded regions indicate the domain Ω, and the spline space spanned by the B-splines over Ω are the same in both cases

Section 4 takes a look at overloading. We look at how to reduce or completely remove overloaded regions in a mesh. We showcase some specific overloading patterns that occur for hierarchical refinement of THB and LRB. Furthermore, we show how local modifications to the LRB-mesh reduce overloading as well as condition numbers.

Section 5 provides a quantitative comparison between the methods. We conduct our numerical experiments using modified central and diagonal refinement examples from [7] with a finer initial tensor-product mesh. This gives enough room on each hierarchical level for the local modifications to take place. The examples show that LRB with no overloading have smaller condition numbers for the mass matrix per degree of freedom than THB and LRB with overloading.

However, the difference seems to be so small that in general all methods have a similar behaviour.

Section 6 summarizes the main results of this paper.

2 Condition Numbers and Knotline Multiplicity at Domain Boundary

In [7], hierarchical refinement was performed for five levels of refinement, using S-THB and S-LRB. The results reported that the number of refinement levels had little to no influence on the evolution of condition numbers for stiffness and mass matrices. There were some minute differences between S-THB and S-LRB, but they followed the same trend. In Fig. 2 we display the condition number of the mass matrix for up to eight refinement levels for S-THB, S-LRB, THB and LRB when run on a hierarchical mesh from [7]. The relevant mesh at the fifth refinement level is displayed in Fig. 4.

The results from [7] is reproduced, and corresponds to the S-THB and S-LRB curves for the first five refinements. However, at the sixth refinement, the curve for the condition number of the mass matrix for S-LRB breaks off and grows exponentially following the curves of LRB that starts three orders of magnitude lower. In Fig. 2 there are also two additional curves (S-LRB1 and LRB1). These are added to show that modifying the mesh by inserting additional knot lines in one parameter direction, with the effect of reducing overloading, significantly reduces the condition numbers of LRB-refinement. This modified mesh is shown in Fig. 4b. We will discuss such modifications more closely in Sect. 4.

Multiplicity of domain boundary knot lines also influences the condition number of the stiffness matrix, as seen in Fig. 3. Here we see that the condition numbers for single boundary knot multiplicity (S-THB, S-LRB and S-LRB1) are two orders of magnitude higher than the condition numbers for open knot vectors (THB, LRB, LRB-1) (Fig. 4).

2.1 Boundary Knotline Multiplicities

We now take a stab at explaining the drastic change in behaviour occuring at $n = 6$ refinements for the S-LRB and S-THB methods as shown in Fig. 2. Since the condition number of a matrix are computed in terms of its largest and smallest eigenvalues, we decided to take a look at the geometric localization of the eigenvectors corresponding to these eigenvalues. By coloring the hierarchical mesh based on the influence of each in terms of the corresponding coefficient in the eigenvector, we obtained a rudimentary geometric visualization of these eigenvectors. In Fig. 5, we see the smallest eigenvector for the mass-matrix corresponding to LRB and S-

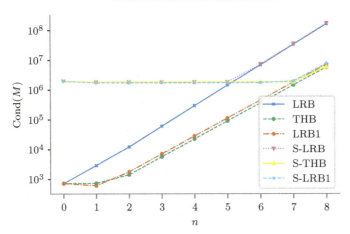

Fig. 2 The condition number of the mass matrix. We see that under repeated refinement, the condition numbers corresponding to spline spaces with open knot vectors (THB, LRB) tends towards the condition numbers corresponding to spline spaces with single knots (S-THB, S-LRB). We also see that a small local modification to reduce overloading in the LRB-space reduces the condition number of the mass matrix (S-LRB1, LRB1)

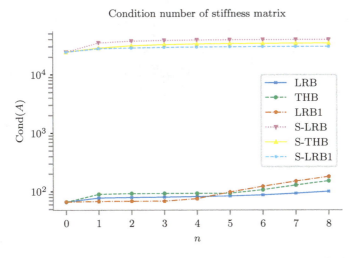

Fig. 3 The condition number of the stiffness matrix. Here the separation between S-LRB, S-THB, LRB and THB are seen in even greater effect

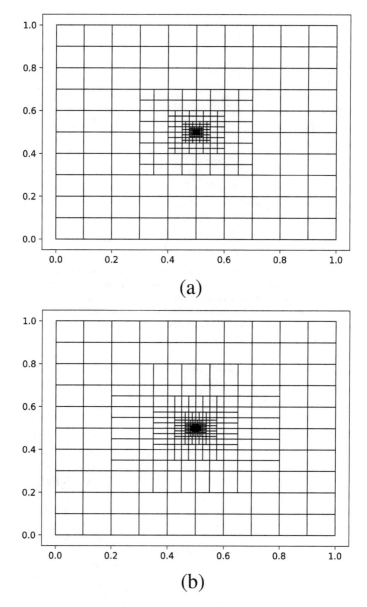

Fig. 4 The meshes used for the preliminary comparison. In (**a**), the unmodified mesh used for S-THB, S-LRB, THB and LRB. In (**b**) the modified mesh used for S-LRB1 and LRB1. This mesh generates a few extra degrees of freedom

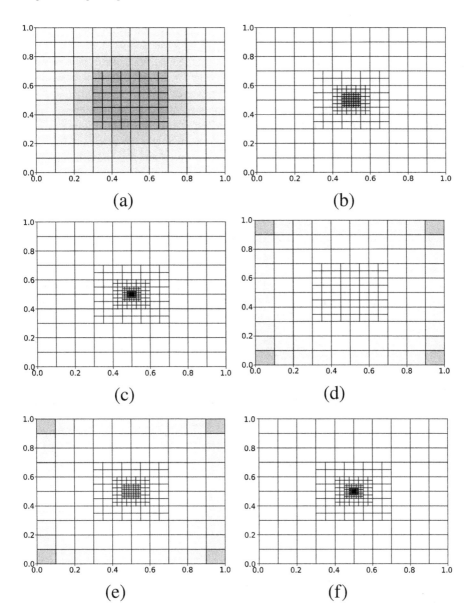

Fig. 5 The eigenvectors corresponding to the smallest eigenvalue of the mass matrix for LRB (**a**) to (**c**), and for S-LRB visualized over the hierarchical mesh after one, three and six refinements (**d**) to (**f**). Darker color indicates higher influence. As we see, the smallest eigenvalues for LRB is localized in the refined region after only one refinement. On the other hand, S-LRB is localized in the corners of the domain up until but not including six refinements, as shown for $n = 1$ and $n = 3$. The effect of the locally refined region dominates only after $n = 6$ refinements as in (**f**). (**a**) $n = 1$ (LRB). (**b**) $n = 3$ (LRB). (**c**) $n = 6$ (LRB). (**d**) $n = 1$ (S-LRB). (**e**) $n = 3$ (S-LRB). (**f**) $n = 6$ (S-LRB)

LRB at the first, third and sixth refinement, and in Fig. 6, the corresponding largest eigenvector.

These figures correspond to the behaviour observed in Fig. 2 where the conditioning for LRB grows after only one refinement, whereas S-LRB needs six refinements before the behaviour in the refined region is registered.

The reason for this behaviour is due to the size of the B-splines defined along the boundary in comparison to the size of the B-splines defined in the interior of the domain. In order to illustrate this, we compute analytically the entries in the mass matrix corresponding to B-splines on various tensor product level and compare these values to the mass matrix entry corresponding to a B-spline defined in the corner of the domain.

2.1.1 Observation for the Mass Matrix

Over the domain $\Omega = [0, 1] \times [0, 1]$ we define a tensor product grid with element size ℓ. In the case of bi-cubic spline spaces, the B-spline defined in the lower left corner of the domain can for LRB and S-LRB be written in terms of their knots as

$$B := B[\mathbf{x}]B[\mathbf{y}],$$
$$Q := B[\mathbf{s}]B[\mathbf{t}], \quad (1)$$

where $\mathbf{x} = \mathbf{y} = [0, 0, 0, 0, \ell]$ and $\mathbf{s} = \mathbf{t} = [-3\ell, -2\ell, -\ell, 0, \ell]$. In both cases, the two B-splines have only one element of support in the domain Ω, namely $\beta := [0, \ell] \times [0, \ell]$. To get a feel for the differences in influence on the mass matrix these B-splines have, we compute the corresponding diagonal elements in the mass matrix.

The polynomial restrictions of B and Q to the element β is

$$B\big|_\beta (x, y) = \frac{(\ell - x)^3 (\ell - y)^3}{\ell^6},$$
$$Q\big|_\beta (x, y) = \frac{(\ell - x)^3 (\ell - y)^3}{36\ell^6}. \quad (2)$$

In other words, $B\big|_\beta = 36 Q\big|_\beta$. If we now compute the diagonal mass matrix entries corresponding to these two elements, we obtain the following:

$$\int_\beta B^2 = \frac{\ell^2}{49}, \quad \int_\beta Q^2 = \frac{\ell^2}{49} \cdot \frac{1}{36^2}. \quad (3)$$

We here see that the matrix element corresponding to the corner B-spline Q arising in S-LRB is three orders of magnitude smaller than the matrix element corresponding to B. Recall the disparity between the curves in Fig. 2, where the differences in the condition numbers also were three orders of magnitude.

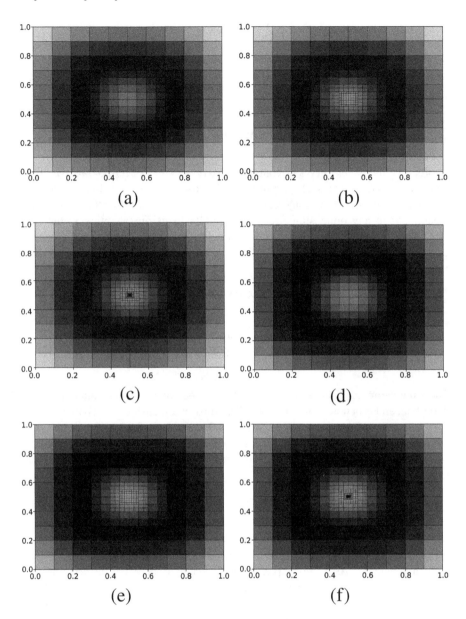

Fig. 6 The eigenvectors corresponding to the largest eigenvalue of the mass matrix for LRB (**a**) to (**c**) and for S-LRB visualized over the hierarchical mesh after one, three and six refinements (**d**) to (**f**). Darker color indicates higher influence. For the largest eigenvalue, the two methods approximately correspond geometrically, and the largest eigenvalues are constant over refinement levels for each of the methods. (**a**) $n = 1$ (LRB). (**b**) $n = 3$ (LRB). (**c**) $n = 6$ (LRB). (**d**) $n = 1$ (S-LRB). (**e**) $n = 3$ (S-LRB). (**f**) $n = 6$ (S-LRB)

Remark 1 This relation between Q and B is both dimension and degree-dependent. The effect will be magnified for higher spatial dimension and higher polynomial degree. A similar condition number analysis was performed for embedded methods in [10], with scaling effects reported along the lines of the effects reported above.

3 Box Partitions, Meshes, and Spline Spaces

In order for the paper to be self-contained, we review the concept of *box partitions* and spline spaces over such partitions in the following section. Readers already familiar with these notions may feel free to skip this section. While the construction generalizes to any dimension, we will gradually focus our attention to the two-dimensional case, as this is most relevant for our discussion. A fully general treatment can be found in [1, 8]. The fundamental building block of a box partition is the *d-dimensional box*.

Definition 1 A **box** β in $\mathbb{R}^d$ (or *d*-**box**) is the Cartesian product of d closed finite intervals $J_1, \ldots, J_d$:

$$\beta = \underset{i=1}{\overset{d}{\times}} J_i. \tag{4}$$

The **dimension** of β is defined to be the number of non-trivial intervals in its definition, and is denoted $\dim(\beta)$. We call a d-box of dimension d an **element**, while a d-box of dimension $d-1$ is called a **mesh-rectangle**. To any mesh-rectangle, we associate an integer k corresponding to which parametric dimension its trivial component resides, and we call the mesh-rectangle a k-**mesh-rectangle** if this has to be emphasized.

In the two-dimensional setting ($d=2$), a **meshline** is a one-dimensional mesh-rectangle.

Remark 2 Note that these naming-conventions are independent of the dimension of the ambient space. Hence, a *mesh-rectangle* may very well be something different from a rectangle. As an example, a mesh-rectangle in $\mathbb{R}^4$ is an axis aligned box. Furthermore, the integer k corresponding to any mesh-rectangle encodes the *direction* of the mesh-rectangle. In the two-dimensional case, where mesh-rectangles are lines, a 1-mesh-rectangle is a vertical line, and a 2-mesh-rectangle is a horizontal line.

As customary in discretizations of computational domains, a large domain is partitioned into a set of non-overlapping smaller geometrical entities. We call such a partition in this specific setting a *box partition*, and this is more precisely defined as follows:

Definition 2 Let $\Omega \subset \mathbb{R}^d$ be an element (d-box of dimension d). A finite collection $\mathcal{E}$ of elements is said to be a **box partition** of Ω if

1. $\beta_i^\circ \cap \beta_j^\circ = \emptyset$ for all $\beta_i, \beta_j \in \mathcal{E}$ where $\beta_i \neq \beta_j$, and
2. $\bigcup_{\beta \in \mathcal{E}} \beta = \Omega$.

In other words, a box partition is an interior-disjoint partition of Ω into a set of smaller elements.

Associated to any element β is its *boundary*, which naturally consists of boxes of dimension one less, i.e., mesh-rectangles. Given a box partition of a larger element Ω, it is therefore sensible to discuss the set of mesh-rectangles associated to this box partition.

Definition 3 (Informal) Given a box partition $\mathcal{E}$ of a domain Ω, we may naturally associate a set of mesh-rectangles $\mathcal{M}$ called a **box mesh** on Ω formed by taking unions of element boundaries.

Remark 3 The link between a box partition $\mathcal{E}$ and the associated box mesh $\mathcal{M}$ is such that by knowing one of them you may recover the other. The box mesh *generated* by a box partition is denoted $\mathcal{M}(\mathcal{E})$, and the box partition *generated* by a box mesh is denoted $\mathcal{E}(\mathcal{M})$.

As our ultimate goal is to define spline spaces based on tensor-product splines over box-partitions, we need to have a concept of knot multiplicity in this more general setting.

Definition 4 A **box mesh with multiplicity** is a pair $(\mathcal{M}, \mu)$ where $\mu\colon \mathcal{M} \to \mathbb{N}$ associates to each mesh-rectangle γ a positive integer $\mu(\gamma)$, called the **multiplicity** of the mesh-rectangle. Note that this is completely analogous to the notion of knot multiplicity for univariate B-splines.

Definition 5 Let a polynomial multi-degree $p = (p_1, \ldots, p_d)$ as well as a box mesh with multiplicity $(\mathcal{M}, \mu)$ corresponding to the box-partition $\mathcal{E}$ of a d-dimensional domain Ω be given. The **spline space of degree p over $\mathcal{M}$** is defined as

$$\mathcal{S}_p^\mu(\mathcal{M}) := \Big\{ f \colon \Omega \to \mathbb{R} : f\big|_\beta \in \Pi_p \text{ for all } \beta \in \mathcal{E}$$

$$\text{and } f \in C^{p_k - \mu(\gamma)} \text{ for all } k\text{-mesh-rectangles } \gamma \in \mathcal{M},$$

$$\text{with } k = 1, \ldots, d \Big\}. \qquad (5)$$

A dimension formula for general spline spaces over box partitions was presented in [9]. In general, the dimension depends on both the topological properties of the box partition and the parametrization of the box partition. In the two-dimensional case—with some requirements on the length of the constituent meshlines—the formula reduces to a formula depending only on the topological features of the mesh. We consider this outside the scope of this text, and refer the reader to [9] for details.

In order to compute with spline spaces over box partitions of the form above, we must be able to construct a set of basis functions that span this space. Several constructions has been studied. We will only be dealing with Truncated Hierarchical B-splines, and Locally Refined B-splines.

Before we move on, we define the notion of a *hierarchical mesh*, a type of box partition over which spline bases of the aforementioned type may be defined. This provides a common ground for comparison of the two methods. The construction is simple and relies on marking regions for which a tensor product mesh of various refinement levels is used.

Definition 6 Let Ω be a domain, and let $\mathcal{M}_1 \subset \cdots \subset \mathcal{M}_M$ be a sequence of nested tensor product meshes on Ω. Let $\Omega_1 \supset \cdots \supset \Omega_M$ be a set of nested subsets of Ω whose boundaries $\partial \Omega_\ell$ align with the meshlines of the corresponding mesh on the coarser level $\mathcal{M}_{\ell-1}$ for $\ell = 2, \ldots, M$. The **hierarchical mesh** $\mathcal{M}$ is defined as

$$\mathcal{M} = \{\gamma \cap \Omega_\ell : \gamma \in \mathcal{M}_\ell \text{ for } \ell = 1, \ldots, M\}, \tag{6}$$

i.e., $\mathcal{M}$ consists of meshlines from each level intersected with the corresponding region, see Fig. 7.

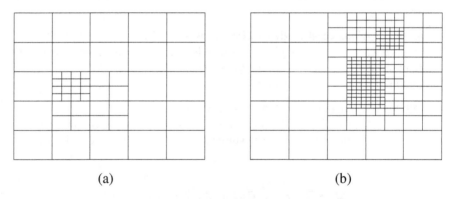

(a)　　　　　　　　　　　　　　(b)

Fig. 7 Two examples of hierarchical meshes. In (**a**) a mesh consisting of three levels of refinement, and in (**b**) a mesh with four levels of refinement. Note here that the region residing at level $\ell = 4$ consists of two disjoint components

3.1 Tensor Product Splines

The foundation for all the locally refined spline spaces over box partitions addressed in this paper, is the tensor product B-spline. We start by glossing over some preliminary definitions.

Recall that a univariate B-spline of polynomial degree p relies on exactly $p+2$ knots. This observation enables us to define B-splines locally without referring to some global knot vector.

Definition 7 Given a polynomial degree p and a non-decreasing knot-vector $\mathbf{t} = (t_1, \ldots, t_{p+2})$, we recursively define the **univariate B-spline** $B[\mathbf{t}]: \mathbb{R} \to \mathbb{R}$ as follows:

If $p = 0$, then

$$B[\mathbf{t}] = \begin{cases} 1, & x \in [t_1, t_2); \\ 0, & \text{otherwise.} \end{cases} \tag{7}$$

If $p > 0$, then

$$B[\mathbf{t}](x) = \frac{x - t_1}{t_{p+1} - t_1} B[\mathbf{t}^-](x) + \frac{t_{p+2} - x}{t_{p+2} - t_2} B[\mathbf{t}^+](x), \tag{8}$$

where $\mathbf{t}^+$ and $\mathbf{t}^-$ are obtained by dropping the first and last elements of $\mathbf{t}$ respectively:

$$\mathbf{t}^+ = (t_2, \ldots, t_{p+2}), \quad \mathbf{t}^- = (t_1, \ldots, t_{p+1}). \tag{9}$$

In the cases of a vanishing denominator, the whole term is taken to be zero.

Such univariate splines can be easily extended to higher dimensions through a tensor product construction.

Definition 8 Let the polynomial multi-degree $p = (p_1, \ldots, p_d)$ and the d local knot vectors $\mathbf{t}^1, \ldots, \mathbf{t}^d$ be given. The d-**variate tensor product B-spline** $B[\mathbf{t}^1, \ldots, \mathbf{t}^d]: \mathbb{R}^d \to \mathbb{R}$ is then defined as

$$B[\mathbf{t}^1, \ldots, \mathbf{t}^d](\mathbf{x}) = \prod_{i=1}^{d} B[\mathbf{t}^i](x_i), \tag{10}$$

where $\mathbf{x} = (x_1, \ldots, x_d)$.

The **support** of $B[\mathbf{t}^1, \ldots, \mathbf{t}^d]$ is the closure of the area where the B-spline takes non-zero values, which we denote by:

$$\text{supp}(B[\mathbf{t}^1, \ldots, \mathbf{t}^d]) = \overline{\left\{\mathbf{x} \in \mathbb{R}^d : B[\mathbf{t}^1, \ldots, \mathbf{t}^d](\mathbf{x}) \neq 0\right\}}. \tag{11}$$

Since our B-spline construction is inherently local, we need to know when a tensor product B-spline has *minimal support* with respect to some box mesh.

Definition 9 (Informal) A B-spline $B = B[\mathbf{t}^1, \ldots, \mathbf{t}^d]$ has **support on** $(\mathcal{M}, \mu)$ if all the knot lines of B occurs as meshlines in $\mathcal{M}$. We say that B has **minimal support on** $(\mathcal{M}, \mu)$ if in addition, all the knot lines of B occur consecutively in $(\mathcal{M}, \mu)$.

One of the central concepts we will be addressing in this paper is the *overloading* of elements. We make this precise in the following definition.

Definition 10 Let a box partition $\mathcal{E}$ of a domain Ω and a polynomial multi-degree $p = (p_1, \ldots, p_d)$ be given. Assume that we construct a set $\mathcal{B}$ of B-splines degree p over the mesh $\mathcal{M}$ corresponding to $\mathcal{E}$. We say that an element β is **overloaded with respect to** $\mathcal{B}$ if the number of B-splines with support on β is larger than the dimension of the corresponding space of polynomials over this element, namely

$$\dim(\Pi_p(\beta)) = \prod_{i=1}^{d}(p_i + 1). \tag{12}$$

We now proceed to review the definitions of LR B-splines and THB-splines.

3.2 Locally Refined Spline Spaces

In preparation for the following discussion, we will adopt the notational convention as in [7] in order to differentiate between the distinct types of basis functions. Depending on the underlying box partition, some of these types may coincide.

Type	Basis	Function
Tensor Product B-spline	$\mathcal{B}$	B
Truncated Hierarchical B-spline	$\mathcal{H}$	H
LR B-spline	$\mathcal{L}$	L

Furthermore, in this and the following sections we will be dealing with box partitions and spline spaces in $\mathbb{R}^2$, unless otherwise stated.

3.2.1 LR-Splines

Locally Refined B-splines (LRB or LR-splines) was introduced by [1]. The LR-spline framework permits the insertion of local splits in a tensor product mesh, and subsequently enables local refinement of the mesh. Being scaled tensor product B-splines, LR-splines admit a set of nice properties. The set of LR B-splines form a partition of unity. Their scaling weights are positive, meaning that they satisfy the convex hull property, and are therefore inherently stable in computations. Moreover, with some restrictions on the refinement process, linear independence of the resulting set of functions can be guaranteed.

LR-splines are defined over so-called *LR-meshes*, being special box partitions. Starting from an initial tensor product mesh, meshlines are inserted sequentially, yielding a sequence of box-meshes, where no meshline is allowed to terminate in the middle of an element. This is formalized in the following definition, and Fig. 8 gives an example.

Definition 11 An **LR mesh** is a box mesh $\mathcal{M} = \mathcal{M}_N$ resulting from a sequence of meshline insertions in an initial tensor product mesh $\mathcal{M}_1$. That is

$$\mathcal{M}_{i+1} = \mathcal{M}_i + \gamma_i \tag{13}$$

for $i = 1, \ldots, N-1$ where each intermediate mesh is a box mesh.

Remark 4 We often think of an LR-mesh as a *sequence* of intermediate meshes

$$\mathcal{M} = \mathcal{M}_N \supseteq \mathcal{M}_{N-1} \supseteq \cdots \supseteq \mathcal{M}_2 \subseteq \mathcal{M}_1 \tag{14}$$

as each intermediate step is needed for the LR B-spline construction.

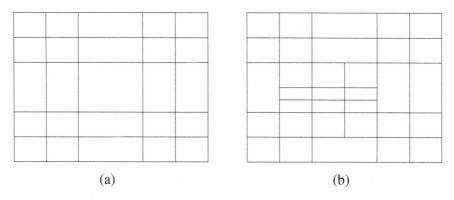

Fig. 8 In (**a**), an initial tensor product mesh, which is also an LR-mesh. In (**b**), an LR-mesh obtained from the insertion of three meshlines in the initial tensor product mesh from (**a**)

Over such an LR-mesh we may define the associated set of LR B-splines algorithmically. Starting from an initial space of tensor product B-splines, meshlines are inserted sequentially. Whenever a meshline completely traverses the support of a B-spline, the B-spline is *split* according to the knot insertion procedure, and two new B-splines are added. The B-spline that was split is removed.

Definition 12 Let $\mathcal{M}$ be an LR-mesh over a domain Ω and $p = (p_1, p_2)$ a polynomial bi-degree. We define the set $\mathcal{L}(\mathcal{M})$ of **LR B-splines of degree** p **over** $\mathcal{M}$ algorithmically as in Algorithm 3.1.

Algorithm 3.1 The LR B-spline construction

Let $\mathcal{L}_1 := \mathcal{B}(\mathcal{M}_1)$ be the set of tensor product B-splines on $\mathcal{M}_1$.
for each intermediate mesh $\mathcal{M}_{i+1} = \mathcal{M}_i + \gamma_i$, with $i = 1, \ldots, N - 1$ **do**
  $\mathcal{L}_{i+1} := \mathcal{L}_i$
  while there exists $B \in \mathcal{L}_{i+1}$ without *minimal support* on $\mathcal{M}_{i+1}$ **do**
    $B^+, B^- = \text{SPLIT}(B, \gamma_i)$ ▷ knotline insertion
    $\mathcal{L}_{i+1} = (\mathcal{L}_{i+1} \setminus \{B\}) \cup \{B^+, B^-\}$ ▷ update the set of B-splines
  end while
end for
$\mathcal{L}(\mathcal{M}) := \mathcal{L}_N$

Remark 5 Note that all LR B-splines have *minimal support* on the resulting mesh. This is by construction. However, there is an important distinction to be made, namely that the set of LR B-splines *differ* from the set of minimal support B-splines on the resulting mesh. This is due to the LR refinement procedure putting some restrictions on the resulting mesh. A survey on the properties of LR-splines and minimal support B-splines are given in [8].

3.2.2 Truncated Hierarchical B-Splines

Hierarchical B-splines, first introduced in [2], is a method for specifying locally refined spline spaces on hierarchical meshes. Recall that a hierarchical mesh consists of regions corresponding to various levels of tensor product grids. The hierarchical B-spline construction involves replacing any B-spline with support completely contained in a region of a finer level by B-splines at this finer level. This procedure will, however, lead to coarse B-splines partially overlapping the finer regions, and does not constitute a partition of unity.

A remedy to this problem came with the introduction of Truncated Hierarchical B-splines [5], where B-splines on a coarse level are *truncated* by B-splines on a finer level. This leads to the resulting set of B-splines forming a partition of unity. The construction relies on the *truncation operator*. Recall that a spline $f \in \text{span}(\mathcal{B}_\ell)$

can be represented in terms of the finer basis $\mathcal{B}_{\ell+1}$:

$$f = \sum_{B_i \in \mathcal{B}_{\ell+1}} c_i^{\ell+1}(f) B_i, \tag{15}$$

where $c_i^{\ell+1}(f)$ is the coefficient multiplying B_i in the representation of f in terms of $\mathcal{B}_{\ell+1}$. For uniform B-splines, this relation is often called the *two-scale relation*. The truncation operator is defined as follows:

Definition 13 Let $B \in \mathcal{B}_\ell$ be a coarse B-spline. The **truncation** with respect to the set of fine B-splines $\mathcal{B}_{\ell+1}$ and the corresponding region $\Omega_{\ell+1}$ is

$$\operatorname{trunc}^{\ell+1} B := \sum_{\substack{B_i \in \mathcal{B}_{\ell+1} \\ \operatorname{supp} B_i \not\subseteq \Omega_{\ell+1}}} c_i^{\ell+1}(B) B_i. \tag{16}$$

Remark 6 The definition above represents the truncation operator in an *additive* sense, where the contributions from the finer level are summed up. It is also possible to represent the truncation operator *subtractively*, by instead removing the bits of the representation that have been replaced by finer B-splines:

$$\operatorname{trunc}^{\ell+1} B = B - \sum_{\substack{B_i \in \mathcal{B}_{\ell+1} \\ \operatorname{supp} B_i \subseteq \Omega_{\ell+1}}} c_i^{\ell+1}(B) B_i \tag{17}$$

Definition 14 Let $\mathcal{M}$ be a hierarchical mesh over a domain Ω (see Definition 6) and $p = (p_1, p_2)$ a polynomial bi-degree. On each level $\ell = 1, \ldots, N$ we have a tensor product spline space V_ℓ spanned by a collection of B-splines $\mathcal{B}_\ell = \mathcal{B}(\mathcal{M}_\ell)$. We define the set of **THB-splines of degree p over $\mathcal{M}$** algorithmically as in Algorithm 3.2.

Algorithm 3.2 The THB-spline construction

Let $\mathcal{H}_1 = \mathcal{B}(\mathcal{M}_1)$ be the set of tensor product B-splines on $\mathcal{M}_1$.
for each level $\ell = 1, \ldots, N-1$ **do**

$$\mathcal{H}_{\ell+1}^{\operatorname{trunc}} := \left\{ \operatorname{trunc}^{\ell+1} H : H \in \mathcal{H}_\ell \text{ and } \operatorname{supp}(H) \not\subseteq \Omega_{\ell+1} \right\}$$

$$\mathcal{H}_{\ell+1}^{\operatorname{new}} := \left\{ B \in \mathcal{B}_{\ell+1} : \operatorname{supp}(\beta) \subseteq \Omega_{\ell+1} \right\}$$

$$\mathcal{H}_{\ell+1} := \mathcal{H}_{\ell+1}^{\operatorname{trunc}} \cup \mathcal{H}_{\ell+1}^{\operatorname{new}}$$

end for
$\mathcal{H} = \mathcal{H}_N$.

Remark 7 Note that in the cases where a B-spline $B \in \mathcal{B}_\ell$ is truncated with respect to $\mathcal{B}_{\ell+1}$ and $\Omega_{\ell+1}$ and the support of B happen to be entirely contained in $\Omega_{\ell+1}$, the truncation operator completely removes the coarse B-spline. In the THB-spline construction, this has the effect of replacing the coarse B-splines with fine B-splines defined in its support.

Remark 8 A simple framework for the implementation of truncated hierarchical B-splines is given in [3], and this serves as a good introduction to the many ways such splines have been implemented in the literature. Efficient algorithms for the assembly of finite element matrices are also presented.

3.2.3 T-Splines

While not directly addressed in this paper, we briefly mention T-splines as LRB with local modifications to the LR-meshes used in this paper will reproduce the spline space generated by semi-standard T-splines [12] and Analysis Suitable T-splines [11]. An example of an Analysis Suitable T-mesh in the index domain is displayed in Fig. 9 to the left, with the corresponding LR-mesh to the right. This is a close up of the structure of a mesh similar to the one used in Fig. 11b.

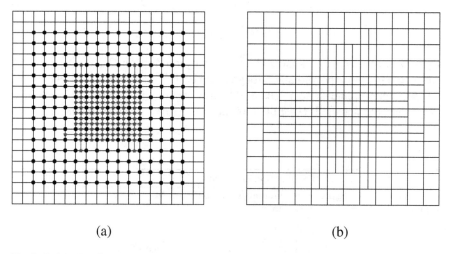

(a)   (b)

Fig. 9 In (**a**) a T-spline mesh in the index domain. The dots denote Greville points or "anchors" for each individual B-spline. A black dot is a B-spline at level $\ell = 0$ and a green star a B-spline at level $\ell = 1$. The resulting spline space can be replicated by an LR-mesh without overloaded elements (c.f. Fig. 11b), as displayed in (**b**). Here we have used multiplicity $m = 4$ along the boundary

4 Local Modification of Meshes and the Reduction of Overloading

In this section we take a deeper look at the overloading of elements, and how local modifications to the mesh may be used to remedy this. Recall from the previous definition that an element β in a box-partition $\mathcal{E}$ is said to be *overloaded* if the number of supported B-splines on the element exceeds the number needed to span the full polynomial space over the element.

We are interested in such overloaded regions, because by reducing or removing completely the overloading on elements we may

1. reduce the bandwidth of the resulting finite element matrices; and
2. improve conditioning of finite element matrices.

Such overloaded regions occur for LRB in convex corners of a fine hierarchical level, where a large B-spline from one hierarchical level overlaps several elements of a finer hierarchical level. For THB, overloading occurs along any border between two hierarchical levels. By coloring in elements with too many supported B-splines we obtain a visualization of this phenomenon, as seen in Fig. 10 on a hierarchical mesh with three levels of refinement.

In order to reduce, or completely remove such overloaded regions, we may for LRB extend meshlines from the fine hierarchical level to the coarse level, in order to split the culprit B-splines. The length needed for this extended meshline depends on the polynomial degree of the B-spline to be split. In Fig. 11 we see the effects of two types of meshline extension to the LRB-mesh from Fig. 10 for a space of bi-cubic splines. The corresponding splines are named LRBNO and T-LRBNO, signifying the fact that these local modifications completely remove overloading.

In order to capture what is happening, we take a closer look at overloading in a convex corner in Fig. 12 where we show how B-splines from the coarse level of a hierarchical mesh may overlap with B-splines from the fine level in such a way that too many B-splines are active over a given element.

5 Numerical Experiments

In order to compare the methods addressed in this paper, we assemble the mass and stiffness matrices associated to discretizations of partial differential equations using IgA or FEM. By computing the condition number of these matrices, we obtain a metric useful for comparison. These matrices arise amongst others in the discretizations of the Poisson equation, and in the computation of the L^2-projection of a function.

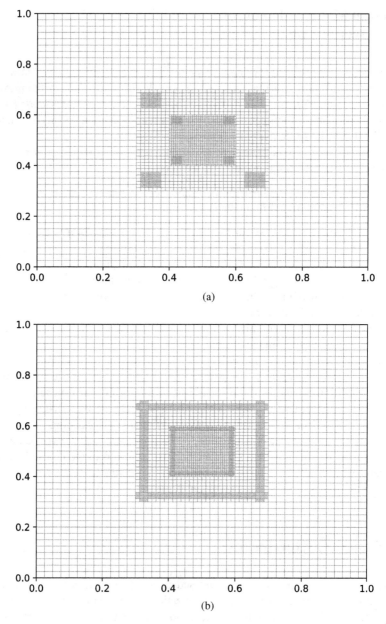

Fig. 10 The overloading patterns on a hierarchical mesh with two levels of refinement. In (**a**), we see that regions in the corners of the refined regions are overloaded, due to the influence of four LR B-splines from the coarser layer, whose support has not been split by any newly introduced meshlines. In (**b**) we observe "bands" of overloaded elements along the boundary between two consecutive refinement levels for THB, arising due to the fact that fine B-splines must be completely contained in the support of a coarse B-spline before truncation occurs. (**a**) LRB. (**b**) THB

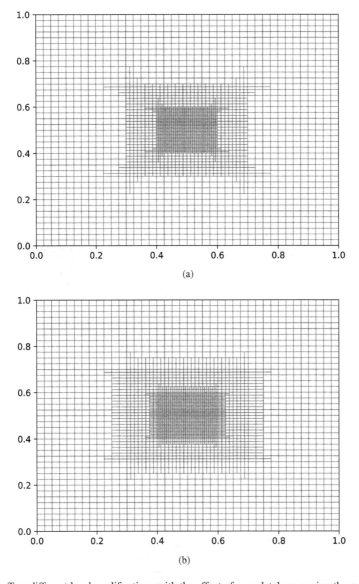

Fig. 11 Two different local modifications with the effect of completely removing the overloaded elements. In (**a**) we extend the meshlines closest to the convex corner by three elements, and the meshline next to them by one element. This has the effect of completely removing the overloading on the corner elements. In (**b**), we make a mesh that can be defined using T-splines that has no overloading. As in (**a**) meshlines closest to the corners are extended by three, while meshlines at the borders between refinement levels are extended by two as in Fig. 9. (**a**) LRBNO. (**b**) T-LRBNO

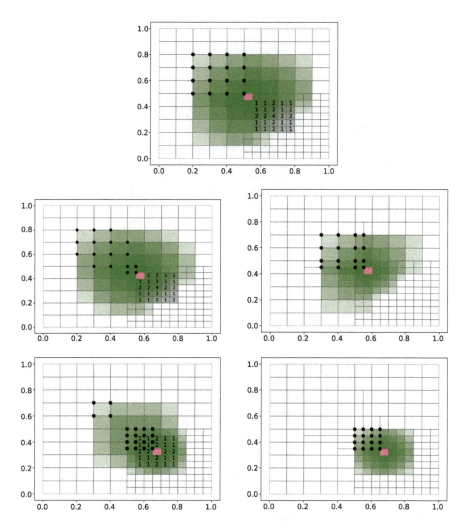

Fig. 12 The effects of extending meshlines on the bi-cubic B-splines covering the element in pink. The upper left corner of each B-spline is marked with a black dot. The knotlines of each B-spline can be identified by starting from the dot and going four knotlines to the right/down. We chose to not use Greville points as some overloaded configurations produce overlapping Greville points. In the *upper mesh* we look at the element just inside the corner of the region refined, and no overloading occurs. In the *middle meshes* we move one element diagonally into the refined region. Before refinement the overloading is one, and after additional lines are inserted the overloading is removed. In the *bottom meshes* we move two additional element diagonally into the refined region, Before refinement the overloading is four, after additional lines are inserted the overloading is removed

5.1 L^2-Projection

Given a domain Ω, a function $f: \Omega \to \mathbb{R}$ in some space of functions V, and a finite-dimensional subspace V_h of V, we are interested in finding the function $u \in V_h$ that minimizes the $L^2(\Omega)$-error

$$\|e\|_{L^2} = \|u - f\|_{L^2}. \tag{18}$$

This can be reformulated as a variational equation by requiring u to satisfy

$$\int_\Omega uv \, d\Omega = \int_\Omega fv \, d\Omega, \tag{19}$$

for all $v \in V_h$. By introducing a basis $\{\varphi_1, \ldots, \varphi_N\}$ for V_h, which in our case will be one of the THB or LRB-bases, we may write this as a linear equation

$$\mathbf{Mc} = \mathbf{b}, \tag{20}$$

where $\mathbf{M}$ is the *mass matrix* and $\mathbf{c}$ is the vector of coefficients representing u in our chosen basis. The entries for $\mathbf{M}$ and the right-hand side $\mathbf{b}$ are given as

$$\mathbf{M}_{ij} = \int_\Omega \varphi_i \varphi_j \, d\Omega, \quad \mathbf{b}_j = \int_\Omega f \varphi_j \, d\Omega. \tag{21}$$

5.2 The Poisson Equation

A commonly encountered differential equation is the *Poisson equation*. Given a function $f: \Omega \to \mathbb{R}$, we wish to find a function u in a space of admissible functions V such that

$$-\Delta u = f \text{ in } \Omega, \tag{22}$$

subject to the boundary conditions

$$u = 0 \text{ on } \Gamma_D, \quad \frac{\partial u}{\partial \mathbf{n}} = g \text{ on } \Gamma_N. \tag{23}$$

Here Γ_D denotes the *Dirichlet*-boundary and Γ_N the *Neumann*-boundary. We assume $\partial\Omega = \Gamma_D \cup \Gamma_N$ and $\Gamma_D \cap \Gamma_N = \emptyset$. Furthermore, $\mathbf{n}$ is the outward facing boundary normal to Ω and g is the prescribed flux along the boundary. This is called the strong form of the Poisson equation.

By multiplying the strong form with a suitable test function, and integrating over the domain, we obtain the variational form of the Poisson equation. The requirements on the smoothness of the sought solution u can be relaxed, by moving some derivatives onto the test-functions. Again, we seek the solution u in a subspace V_h of V spanned by a set of basis functions $\{\varphi_1, \ldots, \varphi_N\}$. The variational form of the Poisson-equation then reads

$$\int_\Omega \nabla u \cdot \nabla v \, d\Omega = \int_{\Gamma_N} gv \, dS - \int_\Omega fv \, d\Omega, \qquad (24)$$

for all $v \in V_h$. Rewriting this in terms of the basis functions, we obtain the system of linear equations

$$\mathbf{Ac} = \mathbf{b}, \qquad (25)$$

where $\mathbf{A}$ is the *stiffness matrix* of the problem. The entries of $\mathbf{A}$ and $\mathbf{b}$ are given as

$$\mathbf{A}_{ij} = \int_\Omega \nabla \varphi_i \cdot \nabla \varphi_j \, d\Omega, \quad \mathbf{b}_j = \int_{\Gamma_N} \varphi_j g \, dS - \int_\Omega f \varphi_j \, d\Omega. \qquad (26)$$

5.3 Condition Numbers

The *condition number* of a matrix $\mathbf{B} \in \mathbb{R}^{n \times n}$ quantifies how sensitive the solution $\mathbf{x}$ to the linear system $\mathbf{Bx} = \mathbf{y}$ is to small perturbations both in $\mathbf{B}$ and the right-hand side $\mathbf{y}$ and is formally defined as

$$\operatorname{Cond}(\mathbf{B}) := \|\mathbf{B}\| \|\mathbf{B}^{-1}\|, \qquad (27)$$

where $\|\cdot\|$ is some matrix norm. Note that the condition number is norm-dependent, but all matrix norms are equivalent on $\mathbb{R}^{n \times n}$. We will be computing the condition numbers in the 2-norm, and in this specific setting for *normal* matrices the condition number can be computed as the ratio between the largest and smallest eigenvalue

$$\operatorname{Cond}(\mathbf{B}) = \frac{|\lambda_1(\mathbf{B})|}{|\lambda_n(\mathbf{B})|}. \qquad (28)$$

Here $\lambda_1 \geq \lambda_2 \geq \ldots \geq \lambda_n$, i.e., ordered in a decreasing fashion.

As in [7], we chose to estimate the condition numbers of the matrices *before* imposing any boundary conditions, as imposing boundary conditions can have a large impact on the conditioning of the matrix. The mass matrix $\mathbf{M}$ is non-singular, even with no imposed boundary condition. The stiffness matrix $\mathbf{A}$ however will be singular, and have a zero-eigenvalue of multiplicity one.

In addition to this, the computation of the smallest eigenvalue of a matrix is a numerically unstable procedure. We will therefore *estimate* the condition numbers as follows:

$$\mathrm{Cond}(\mathbf{M}) \approx \frac{|\lambda_1(\mathbf{M})|}{|\lambda_n(\mathbf{M})|}, \text{ and } \mathrm{Cond}(\mathbf{A}) \approx \frac{|\lambda_1(\mathbf{A})|}{|\lambda_{n-1}(\mathbf{A})|}, \tag{29}$$

using the second-smallest eigenvalue for the stiffness matrix.

5.4 Numerical Results

Below we present results of the numerical simulations. As LRBNO generates higher dimensional spline spaces than THB and LRB we plot the condition numbers as a function of the degrees of freedom. Just plotting the condition numbers as a function of the levels provides less information. By including the dimension of the spline space we obtain a clearer distinction between the methods.

5.4.1 Central Refinement

We assemble the stiffness and mass matrices on a sequence of meshes corresponding to central refinement, shown at the third refinement for bi-cubic splines in Figs. 10 and 11. In addition to the bi-cubic case, we also assemble the matrices on similar meshes for bi-quadratic and bi-quartic spline spaces, where the spacing between each refined region is kept the same for all degrees. The results are shown in Figs. 13 and 14. Unfortunately, due to time constraints, we were not able to get results for bi-quartic THB-splines.

Start by noting that for the mass matrix, THB performs better than LRB with no modifications, while for the stiffness matrices, the two methods are comparable with LRB having a slight advantage. The number of degrees of freedom are the same. By locally modifying the mesh, as is the case for LRBNO and T-LRBNO, we see that the number of degrees of freedom goes up, as expected. The condition number per degree of freedom is smallest for T-LRBNO.

5.4.2 Diagonal Refinements

We now consider the case of diagonal refinement for bi-cubic spline spaces. Again, we use the same hierarchical mesh for LRB and THB. We will only consider one sequence of meshes with local modifications. In the diagonal refinement setting, the corners of the refined region are sufficiently close to each other so that we need to make a decision on which direction to refine in. The diagonally refined mesh is not compatible with a T-spline type mesh, and will therefore not be taken into consideration here.

We assemble stiffness and mass matrices on the meshes displayed in Fig. 15. The results are shown in Figs. 16 and 17. Note that for the diagonal refinement, the number of degrees of freedom generated when removing overloading, shown in

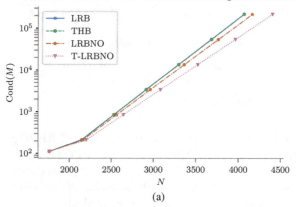

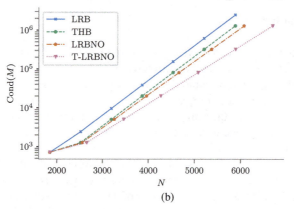

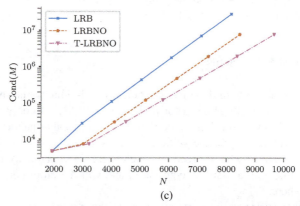

Fig. 13 The condition numbers for mass matrices over a centrally refined hierarchical mesh for six levels of refinement. In the figures N denotes the number of degrees of freedom in the corresponding space. (**a**) Quadratic. (**b**) Cubic. (**c**) Quartic

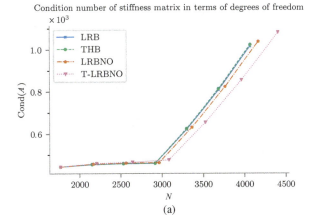

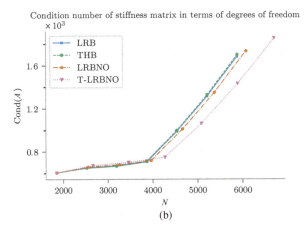

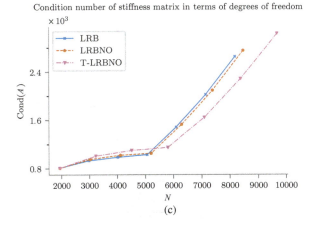

Fig. 14 The condition numbers for stiffness matrices over a centrally refined hierarchical mesh for six levels of refinement. In the figures N denotes the number of degrees of freedom in the corresponding space. (**a**) Quadratic. (**b**) Cubic. (**c**) Quartic

Fig. 15 The overloading patterns on a hierarchical mesh with three levels of diagonal refinement. In this case, we see in greater effect the behaviour of LRB over convex corners. Here the difference in overloading between THB and LRB are smaller, as opposed to the central refinement case, due to the high number of corners relative to the length of the sides of the refined levels. By using a one-directional meshline extension along the diagonal, and extensions similar to the central-refinement case, we may completely remove overloading. (**a**) LRB. (**b**) THB. (**c**) LRBNO

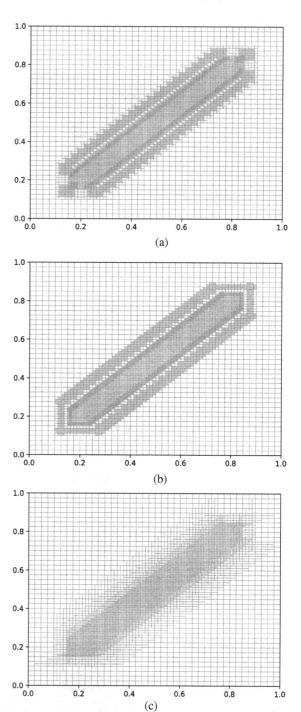

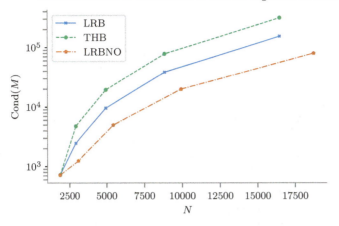

Fig. 16 Condition numbers for mass matrices over a diagonally refined hierarchical mesh for four levels of refinement. There is one data point for each method at each refinement level. The first point is the same for all methods as the all methods start from the same tensor product spline space. N denotes the number of degrees of freedom in the corresponding spline space. The none overloaded LRBNO mesh has clearly the smallest condition numbers

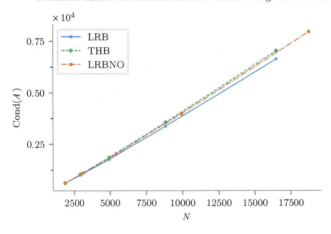

Fig. 17 The condition numbers for stiffness matrices over a diagonally refined hierarchical mesh for four levels of refinement. In the figures N denotes the number of degrees of freedom in the corresponding space. All methods are similar in behaviour with respect to condition numbers as a function of degrees of freedom

the mesh in Fig. 15c, is a fair bit larger than the unmodified counterparts. Despite this, LRBNO outperforms THB and LRB by a significant amount when it comes to the mass matrix. The conditioning of the stiffness matrix on the other hand grows approximately linearly with the number of degrees of freedom, and no significant effect of the overload-reduction can be seen.

6 Conclusion

We have addressed differences and similarities of Truncated Hierarchical B-splines (THB) and Locally Refined B-splines (LRB) on similar hierarchical meshes. The overall conclusion is that there are no big differences between the methods with respect to condition numbers of mass and stiffness matrices for the example meshes addressed.

- When THB and LRB are run on identical meshes THB has better condition numbers for the mass matrix except for the most complex example run, the diagonal example in Figs. 15 and 16. The behaviour of the stiffness matrix is very similar for both methods.
- When making a mesh for LRB that has no overloading the condition numbers for the mass matrix of LRB are smaller than those of THB, with condition numbers of stiffness being similar. It should be noted that using meshes for LRB that has no overloading guarantees that the B-splines generated are linearly independent, and that the number of B-splines covering an element is the minimal needed for spanning the polynomial space over the element. For hierarchical meshes of bi-degree less than $(4, 4)$ there is always linear independence in the set of LR B-splines generated. For bi-degree $(4, 4)$ and higher linear dependence can occur in very special configurations when the elements outside two opposing concave corners of a refinement region is covered by the same B-spline from a cruder level. This happens for bi-degree $(4,4)$ when a refinement region is split if just one element from the cruder level is not refined, e.g., the refinement region is locally very narrow.

When trying to represent hierarchical refinements using T-splines as in Fig. 9 there is a region of one directional refinement of length two just outside the boundary of the refinement region. This gives a smoother transition between refinement levels that can also be replicated by LRB. The results in Figs. 13 and 14 show a better behaviour than going directly from one refinement level to the next. Having such an intermediate level of refinement if possible is advantageous. However, in situations such as the diagonal refinement in Fig. 15 this is not possible.

Most often THB is described as based on dyadic sequences of grids determined by scaled lattices over which uniform B-spline spaces are defined. This implies that there is single knot multiplicity along domain boundaries. However, variants of THB are published [4] where open knots are used along the domain boundary. In Sect. 2 we have shown that open knot vectors are preferable, not only with respect to simplified interpolation of boundary conditions, but also to avoid that the condition number of the mass matrix is biased by the boundary B-splines. As we see the same effect for LR B-splines we have a strong recommendation that open knot vectors are used for locally refined splines, rather than single knot multiplicity at domain boundaries.

Acknowledgments This project has received funding from the The Research Council of Norway under grant agreement No 270922.

References

1. Tor Dokken, Tom Lyche, and Kjell Fredrik Pettersen. "Polynomial Splines over Locally Refined Box-Partitions". In: *Computer Aided Geometric Design* 30.3 (Mar. 2013), pp. 331–356.
2. David R. Forsey and Richard H. Bartels. "Hierarchical B-Spline Refinement". In: *Proceedings of the 15th Annual Conference on Computer Graphics and Interactive Techniques*. SIGGRAPH '88. New York, NY, USA: ACM, pp. 205–212.
3. Eduardo M. Garau and Rafael Vázquez. "Algorithms for the Implementation of Adaptive Isogeometric Methods Using Hierarchical B-Splines". In: *Applied Numerical Mathematics* 123 (Jan. 1, 2018), pp. 58–87.
4. Carlotta Giannelli, Bert Jüttler, Stefan K. Kleiss, Angelos Mantzaflaris, Bernd Simeon, and Jaka Špeh. "Thb-splines: An effective mathematical technology for adaptive refinement in geometric design and isogeometric analysis". In: *Computer Methods in Applied Mechanics and Engineering* 299 (2016), pp. 337–365.
5. Carlotta Giannelli, Bert Juttler, and Hendrik Speleers. "THB-Splines: The Truncated Basis for Hierarchical Splines". In: *Computer Aided Geometric Design* 29.7 (Oct. 1, 2012), pp. 485–498.
6. Kjetil André Johannessen, Trond Kvamsdal, and Tor Dokken. "Isogeometric Analysis Using LR B-Splines". In: *Computer Methods in Applied Mechanics and Engineering* 269 (Feb. 2014), pp. 471–514.
7. Kjetil André Johannessen, Filippo Remonato, and Trond Kvamsdal. "On the Similarities and Differences between Classical Hierarchical, Truncated Hierarchical and LR B-Splines". In: *Computer Methods in Applied Mechanics and Engineering* 291 (July 1, 2015), pp. 64–101.
8. Francesco Patrizi and Tor Dokken. "Linear dependence of bivariate Minimal Support and LR B-Splines over LR-Meshes". In: (Nov. 12, 2018). arXiv: 1811.04919 [math].
9. Kjell Fredrik Pettersen. *On the Dimension of Multivariate Spline Spaces*. Tech. rep. http://hdl.handle.net/11250/2432305. 2013.
10. F. de Prenter, C.V. Verhoosel, G.J. van Zwieten, and E.H. van Brummelen. "Condition number analysis and preconditioning of the finite cell method". In: *Computer Methods in Applied Mechanics and Engineering* 316(2017). Special Issue on Isogeometric Analysis: Progress and Challenges, pp. 297–327.
11. M.A. Scott, X. Li, T.W. Sederberg, and T.J.R. Hughes. "Local refinement of analysis-suitable T-splines". In: *Computer Methods in Applied Mechanics and Engineering* 213–216 (2012), pp. 206–222.
12. Thomas W. Sederberg, David L. Cardon, G. Thomas Finnigan, Nicholas S. North, Jianmin Zheng, and Tom Lyche. "T-spline Simplification and Local Refinement". In: *ACM Trans. Graph.* 23.3 (Aug. 2004), pp. 276–283.
13. A.-V. Vuong, C. Giannelli, B. Jüttler, and B. Simeon. "A hierarchical approach to adaptive local refinement in isogeometric analysis". In: *Computer Methods in Applied Mechanics and Engineering* 200.49 (2011), pp. 3554–3567.

Efficient p-Multigrid Based Solvers for Isogeometric Analysis on Multipatch Geometries

Roel Tielen, Matthias Möller, and Cornelis Vuik

Abstract Isogeometric Analysis can be considered as the natural extension of the Finite Element Method (FEM) to higher-order spline based discretizations simplifying the treatment of complex geometries with curved boundaries. Finding a solution of the resulting linear systems of equations efficiently remains, however, a challenging task. Recently, p-multigrid methods have been considered [18], in which a multigrid hierarchy is constructed based on different approximation orders p instead of mesh widths h as it would be the case in classical h-multigrid schemes [8]. The use of an Incomplete LU-factorization as a smoother within the p-multigrid method has shown to lead to convergence rates independent of both h and p for single patch geometries [19]. In this paper, the focus lies on the application of the aforementioned p-multigrid method on multipatch geometries having a C^0-continuous coupling between the patches. The use of ILUT as a smoother within p-multigrid methods leads to convergence rates that are essentially independent of h and p, but depend mildly on the number of patches.

1 Introduction

Isogeometric Analysis (IgA) [9] can be considered as the natural extension of the Finite Element Method (FEM) to higher-order spline based discretizations simplifying the treatment of complex geometries with curved boundaries. However, solving the resulting linear systems arising in IgA efficiently is considered a challenging task, especially for higher-order discretizations. The exponential increase of the condition numbers of the mass and stiffness matrices with the approximation order p, make the use of (standard) iterative solvers inefficient. The wider support of the

---

R. Tielen (✉) · M. Möller · C. Vuik
Delft Institute of Applied Mathematics, Delft University of Technology, Delft, The Netherlands
e-mail: r.p.w.m.tielen@tudelft.nl

basis functions and, consequently, increasing bandwidth of the matrices for larger values of p make the use of direct solvers on the other hand also not straightforward.

The use of established solution techniques from FEM in IgA has been an active field of research. For example, h-multigrid methods have been investigated, as they are considered among the most efficient solvers in Finite Element Methods for elliptic problems. The use of standard smoothers like (damped) Jacobi or Gauss-Seidel leads, however, to convergence rates which deteriorate drastically for increasing values of p [5], caused by very small eigenvalues associated with high-frequency eigenvectors [3]. Non-classical smoothers have been developed to solve this problem leading to (geometric) multigrid methods which are robust in both h and p [8, 13].

An alternative solution strategy are p-multigrid methods. In contrast to h-multigrid methods, where each level of the constructed hierarchy is obtained by refining the mesh, in p-multigrid methods each level represents a different approximation order. p-Multigrid methods are widely used within the Discontinuous Galerkin framework [4, 10, 11, 14], where $p = 0$ is used on the coarsest hierarchy level.

Some research has been performed for continuous Galerkin methods [7] as well, where the coarse grid correction was obtained at level $p = 1$. Throughout this paper, the coarse grid is also obtained at level $p = 1$. Here, B-spline basis functions coincide with piecewise-linear ($p = 1$) Lagrange basis functions, enabling the use of well known solution techniques for standard FEM.

Recently, the authors developed an efficient p-multigrid method for IgA discretizations that makes use of an Incomplete LU factorization based on a dual threshold strategy (ILUT) [15] as a smoother. This approach was shown to result in a p-multigrid method with essentially h- and p-independent convergence rates [19] in contrast to the use of Gauss-Seidel as a smoother.

In this paper, the focus lies on the extension of p-multigrid based methods on multipatch geometries, giving rise to (reduced) C^0-continuity between individual patches. The spectral properties of the p-multigrid method are analysed and numerical results are presented for different two-dimensional benchmarks. The use of ILUT as a smoother leads to a p-multigrid method that shows essentially h- and p-independent convergence rates on multipatch geometries. Furthermore, the number of iterations needed to achieve convergence depends only mildly on the number of patches.

This paper is organised as follows. The model problem and spatial discretization are briefly considered in Sect. 2. Section 3 presents the p-multigrid method together with the adopted ILUT smoother in more detail. In Sect. 4, a spectral analysis is performed and discussed. Numerical results for the considered benchmarks are presented in Sect. 5. Conclusions are finally drawn in Sect. 6.

2 Model Problem

As a model problem to describe the spatial discretisation, Poisson's equation is considered:

$$-\Delta u = f, \quad \text{on } \Omega, \tag{1}$$

where $\Omega \subset \mathbb{R}^2$ is a connected, Lipschitz domain, $f \in L^2(\Omega)$ and $u = 0$ on the boundary $\partial \Omega$. Let $\mathcal{V} = H_0^1(\Omega)$ denote the subspace of the Sobolev space $H^1(\Omega)$ that contains all functions that vanish on the boundary $\partial \Omega$. By multiplying both sides of (1) with an arbitrary test function $v \in \mathcal{V}$ and applying integration by parts in the left side, the following variational form of (1) is obtained:

Find $u \in \mathcal{V}$ such that

$$a(u, v) = (f, v) \quad \forall v \in \mathcal{V}, \tag{2}$$

where

$$a(u, v) = \int_\Omega \nabla u \cdot \nabla v \, d\Omega \quad \text{and} \quad (f, v) = \int_\Omega f v \, d\Omega. \tag{3}$$

A bijective geometry function $\mathbf{F}$ is then defined to parameterize the physical domain Ω:

$$\mathbf{F} : \Omega_0 \to \Omega, \qquad \mathbf{F}(\boldsymbol{\xi}) = \mathbf{x}, \tag{4}$$

where $\boldsymbol{\xi} = (\xi, \eta)$ and $\mathbf{x} = (x, y)$ denote the coordinates in the parametric and physical domain, respectively. The geometry function $\mathbf{F}$ describes an invertible mapping connecting the parameter domain $\Omega_0 \subset \mathbb{R}^2$ with the physical domain Ω. In case Ω cannot be described by a single geometry function, the physical domain is divided in a collection of non-overlapping subdomains $\Omega^{(k)}$ such that

$$\Omega = \bigcup_{k=1}^{K} \Omega^{(k)}. \tag{5}$$

A geometry function $\mathbf{F}^{(k)}$ is then defined to parameterize each subdomain $\Omega^{(k)}$:

$$\mathbf{F}^{(k)} : \Omega_0 \to \Omega^{(k)}, \qquad \mathbf{F}^{(k)}(\boldsymbol{\xi}) = \mathbf{x}. \tag{6}$$

We refer to Ω as a multipatch geometry consisting of K patches. Throughout this paper, the tensor product of one-dimensional B-spline basis functions $\phi_{i_x,p}(\xi)$ and $\phi_{i_y,q}(\eta)$ of order p and q, respectively, with maximum continuity are adopted for the spatial discretisation:

$$\Phi_{\mathbf{i},\mathbf{p}}(\boldsymbol{\xi}) := \phi_{i_x,p}(\xi)\phi_{i_y,q}(\eta), \qquad \mathbf{i} = (i_x, i_y), \; \mathbf{p} = (p, q). \tag{7}$$

Here, $\mathbf{i}$ and $\mathbf{p}$ are multi indices, with $i_x = 1, \ldots, n_x$ and $i_y = 1, \ldots, n_y$ denoting the one-dimensional basis functions in the x and y-dimension, respectively. Furthermore, $i = i_x n_x + i_y n_y$ assigns a unique index to each pair of one-dimensional basis functions, where $i = 1, \ldots N_{\text{dof}}$. The spline space $\mathcal{V}_{h,p}$ can then be written, using the inverse of the geometry mapping $\mathbf{F}^{-1}$ as pull-back operator, as follows:

$$\mathcal{V}_{h,p} := \text{span}\left\{\Phi_{\mathbf{i},\mathbf{p}} \circ \mathbf{F}^{-1}\right\}_{i=1,\ldots,N_{\text{dof}}}. \tag{8}$$

Here, N_{dof} denotes the number of degrees of freedom, or equivalently, the number of tensor-product basis functions. The Galerkin formulation of (2) becomes:

Find $u_{h,p} \in \mathcal{V}_{h,p}$ such that

$$a(\nabla u_{h,p}, \nabla v_{h,p}) = (f, v_{h,p}) \qquad \forall v_{h,p} \in \mathcal{V}_{h,p}, \tag{9}$$

or, equivalently:

$$\mathbf{A}_{h,p}\mathbf{u}_{h,p} = \mathbf{f}_{h,p}. \tag{10}$$

Here, $\mathbf{A}_{h,p}$ denotes the stiffness matrix resulting from the discretization of the left-hand side with the tensor-product of B-spline basis functions of order p and knot span size h. To assess the quality of the p-multigrid method throughout this paper, the following benchmarks are considered:

Benchmark 1 Here, Poisson's equation is considered on the unit square, i.e. $\Omega = [0, 1]^2$. The right-hand side is chosen such that the exact solution u is given by:

$$u(x, y) = \sin(\pi x)\sin(\pi y).$$

Benchmark 2 Let Ω be the quarter annulus with an inner and outer radius of 1 and 2, respectively. Again, Poisson's equation is considered, where the exact solution u is given by

$$u(x, y) = -(x^2 + y^2 - 1)(x^2 + y^2 - 4)xy^2,$$

Benchmark 3 Let $\Omega = \{[-1, 1] \times [-1, 1]\} \setminus \{[0, 1] \times [0, 1]\}$ be an L-shaped domain. As with the other benchmarks, Poisson's equation is considered, where the exact solution u is given by

$$u(x, y) = \begin{cases} \sqrt[3]{x^2 + y^2}\sin\left(\frac{2\text{atan2}(y,x)-\pi}{3}\right) & \text{if } y > 0 \\ \sqrt[3]{x^2 + y^2}\sin\left(\frac{2\text{atan2}(y,x)+3\pi}{3}\right) & \text{if } y < 0 \end{cases},$$

where atan2 is the 2-argument arctangent function. The right-hand side is chosen according to the exact solution. For the first two benchmarks, homogeneous Dirichlet boundary conditions are applied on the entire boundary $\partial\Omega$, while for the third benchmark inhomogeneous Dirichlet boundary conditions are applied. Note that the geometry of each benchmark can be described by a single patch. The multipatch geometries considered throughout this paper are obtained by splitting the single patch uniformly in both directions.

3 p-Multigrid Method

Multigrid methods solve linear systems of equations by defining a hierarchy of discretizations. At each level of the hierarchy, a basic iterative method, like Gauss-Seidel or (damped) Jacobi, is then applied as a smoother. On the coarsest level, a correction is determined by solving the residual equation. With p-multigrid methods, a sequence of spaces $\mathcal{V}_{h,1}, \ldots, \mathcal{V}_{h,p}$ is obtained by applying refinement in p. The coarse grid correction is then determined at level $p = 1$. Since basis functions with maximal continuity are considered, the spaces in the hierarchy are not nested. For p-multigrid, the two-grid correction scheme consists of the following steps [18, 19]:

1. Apply a fixed number ν_1 of presmoothing steps to update the initial guess $\mathbf{u}_{h,p}^{(0)}$:

$$\mathbf{u}_{h,p}^{(0,m)} = \mathbf{u}_{h,p}^{(0,m-1)} + \mathcal{S}_{h,p}\left(\mathbf{f}_{h,p} - \mathbf{A}_{h,p}\mathbf{u}_{h,p}^{(0,m-1)}\right), \quad m = 1, \ldots, \nu_1. \quad (11)$$

Here, $\mathcal{S}$ is a smoothing operator applied to the high-order problem.

2. Project the residual at level p onto $\mathcal{V}_{h,p-1}$ using the restriction operator $\mathcal{I}_p^{p-1}$:

$$\mathbf{r}_{h,p-1} = \mathcal{I}_p^{p-1}\left(\mathbf{f}_{h,p} - \mathbf{A}_{h,p}\mathbf{u}_{h,p}^{(0,\nu_1)}\right). \quad (12)$$

3. Determine the coarse grid error, by solving the residual equation at level $p - 1$:

$$\mathbf{A}_{h,p-1}\mathbf{e}_{h,p-1} = \mathbf{r}_{h,p-1}. \quad (13)$$

4. Use the prolongation operator $\mathcal{I}_{p-1}^p$ to project the error $\mathbf{e}_{h,p-1}$ onto the space $\mathcal{V}_{h,p}$ and update $\mathbf{u}_{h,p}^{(0,\nu_1)}$:

$$\mathbf{u}_{h,p}^{(0,\nu_1)} := \mathbf{u}_{h,p}^{(0,\nu_1)} + \mathcal{I}_{p-1}^p\left(\mathbf{e}_{h,p-1}\right). \quad (14)$$

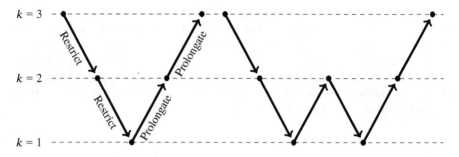

Fig. 1 Description of a V-cycle and W-cycle

5. Apply ν_2 postsmoothing steps to obtain $\mathbf{u}_{h,p}^{(0,\nu_1+\nu_2)} =: \mathbf{u}_{h,p}^{(1)}$:

$$\mathbf{u}_{h,p}^{(0,\nu_1+m)} = \mathbf{u}_{h,p}^{(0,\nu_1+m-1)} + \mathcal{S}_{h,p}\left(\mathbf{f}_{h,p} - \mathbf{A}_{h,p}\mathbf{u}_{h,p}^{(0,\nu_1+m-1)}\right), \quad m = 1, \ldots, \nu_2. \quad (15)$$

The residual equation can be solved recursively by applying the same two-grid correction scheme until level $p = 1$, which results in a V-cycle. Different cycle types can be applied, however, as shown in Fig. 1.

3.1 Prolongation and Restriction

The prolongation and restriction operators transfer both coarse grid corrections and residuals between different levels of the hierarchy. The prolongation and restriction operator adopted in this paper are based on the L_2 projection [1, 2, 17]. The prolongation operator $\mathcal{I}_{k-1}^k : \mathcal{V}_{h,k-1} \to \mathcal{V}_{h,k}$ can be derived from the following variational form

$$(\mathcal{I}_{k-1}^k(u_{h,k-1}), v_{h,k}) = (u_{h,k-1}, v_{h,k}) \quad \forall u_{h,k-1} \in \mathcal{V}_{k-1}, \quad \forall v_{h,k} \in \mathcal{V}_k \quad (16)$$

and is therefore given by

$$\mathcal{I}_{k-1}^k(\mathbf{v}_{k-1}) = (\mathbf{M}_k)^{-1} \mathbf{P}_{k-1}^k \mathbf{v}_{k-1}, \quad (17)$$

where the mass matrix $\mathbf{M}_k$ and the transfer matrix $\mathbf{P}_{k-1}^k$ are defined, respectively, as follows:

$$(\mathbf{M}_k)_{(i,j)} := \int_\Omega \Phi_{\mathbf{i},\mathbf{k}} \Phi_{\mathbf{j},\mathbf{k}} \, d\Omega, \qquad (\mathbf{P}_{k-1}^k)_{(i,j)} := \int_\Omega \Phi_{\mathbf{i},\mathbf{k}} \Phi_{\mathbf{j},\mathbf{k-1}} \, d\Omega. \quad (18)$$

The restriction operator $I_k^{k-1} : \mathcal{V}_{h,k} \to \mathcal{V}_{h,k-1}$ is given by the Hilbert adjoint of the prolongation operator and defined by

$$I_k^{k-1}(\mathbf{v}_k) = (\mathbf{M}_{k-1})^{-1} \mathbf{P}_k^{k-1} \mathbf{v}_k. \tag{19}$$

The explicit solution of a linear system of equations for each projection step is prevented by replacing the consistent mass matrix $\mathbf{M}$ in both transfer operators by its lumped counterpart $\mathbf{M}^L$. Here, $\mathbf{M}^L$ is obtained by applying row-sum lumping:

$$\mathbf{M}^L_{(i,i)} = \sum_{j=1}^{N_{\text{dof}}} \mathbf{M}_{(i,j)}. \tag{20}$$

3.2 Smoother

In this paper, an Incomplete LU factorization with a dual threshold strategy (ILUT) [15] is adopted as a smoother. The ILUT factorization is completely determined by a tolerance τ and fillfactor f. All matrix entries in the factorization smaller (in absolute value) than the tolerance multiplied by the average magnitude of all entries in the current row are dropped. Furthermore, only the average number of non-zeros in each row of the original operator $\mathbf{A}_{h,p}$ multiplied with the fillfactor are kept in each row.

Throughout this paper, a fillfactor of 1 is adopted and the dropping tolerance τ equals 10^{-12}. As a consequence, the number of non-zero entries of the factorization is similar to the number of non-zeros of the original operator. An efficient implementation of an ILUT factorization is available in the Eigen library [6] based on [16]. Once the factorization is obtained, a single smoothing step is applied as follows:

$$\mathbf{e}_{h,p}^{(n)} = (\mathbf{L}_{h,p} \mathbf{U}_{h,p})^{-1} (\mathbf{f}_{h,p} - \mathbf{A}_{h,p} \mathbf{u}_{h,p}^{(n)}), \tag{21}$$

$$\mathbf{u}_{h,p}^{(n+1)} = \mathbf{u}_{h,p}^{(n)} + \mathbf{e}_{h,p}^{(n)}. \tag{22}$$

3.3 Coarse Grid Operator

The system operator $\mathbf{A}_{h,p}$ is needed at each level of the hierarchy to apply the smoothing steps or solve the residual equation at level $p = 1$. The operators at the coarser levels can be obtained by rediscretizing the bilinear form in (9) with lower-order spline basis functions or by applying a Galerkin projection:

$$\mathbf{A}_{h,k-1}^G = I_k^{k-1} \mathbf{A}_{h,k} I_{k-1}^k. \tag{23}$$

4 Spectral Analysis

To investigate the performance of the p-multigrid method on multipatch geometries, the spectrum of the iteration matrix is determined. The iteration matrix describes the effect of a single multigrid cycle on $\mathbf{u}_{h,p}$ and can be used to obtain the asymptotic convergence rate of the p-multigrid method. For all benchmarks introduced in Sect. 2, results are presented considering a different number of patches.

The asymptotic convergence rate of a multigrid method is determined by the spectral radius of the corresponding iteration matrix. This matrix can be obtained explicitly by considering $-\Delta u = 0$ with homogeneous Dirichlet boundary conditions. By applying a single iteration of the p-multigrid method using the ith unit vector as initial guess, one obtains the ith column of the iteration matrix [20].

The spectra obtained for the first two benchmarks are shown in Fig. 2 for a different number of patches. The multipatch geometries are obtained by splitting the single patch uniformly in both directions, leading to 4 or 16 patches with a C^0-continuous coupling at the interfaces. For the single patch, all eigenvalues of the iteration matrix are clustered around the origin. For the multipatch geometries, some eigenvalues are slightly further from the origin. Table 1, showing the spectral radius of the iteration matrix for different values of h and p for the first benchmark, confirms this observation. The spectral radii are larger for all numerical experiments when the number of patches is increased, but still relatively low. Furthermore, since the spectral radii remain almost constant for higher values of p, the p-multigrid method is expected to show (essentially) p-independent convergence rates.

The obtained spectral radii for the second benchmark for different values of h and p can be found in Table 2. Again, the multipatch geometries consist of 4 and 16 patches. For all configurations, the spectral radius for a single patch geometry is lower compared to the spectral radius obtained for the multipatch geometries. As a consequence, the p-multigrid is expected to show slower convergence behaviour for multipatch geometries. On the other hand, the asymptotic convergence rates for the multipatch geometries are almost independent of p and still relatively low. For a single configuration the resulting p-multigrid method is unstable, which is reflected by a spectral radius larger then 1. The obtained spectral radii for the third benchmark are presented in Table 3. As with the other benchmarks, the spectral radii remain almost constant for higher values of p, implying (essentially) p-independent convergence rates.

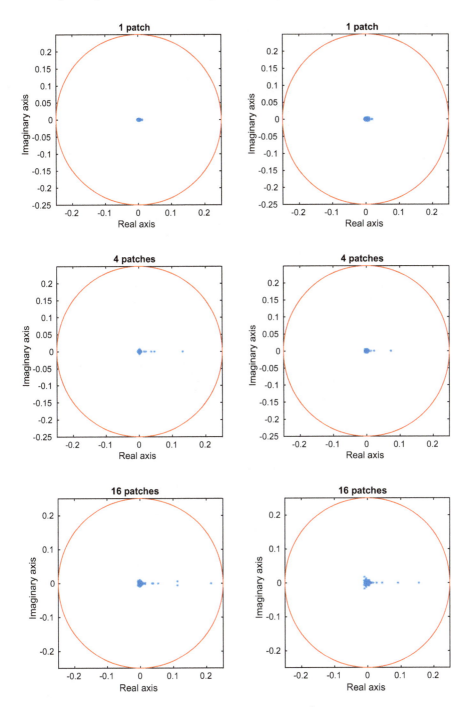

Fig. 2 Spectra of the iteration matrix (with $p = 3$ and $h = 2^{-5}$) for the first (left) and second (right) benchmark for a different number of patches

Table 1 Spectral radius ρ for the first benchmark for different values of h and p for a different number of patches

	# patches				# patches				# patches		
$p=2$	1	4	16	$p=3$	1	4	16	$p=4$	1	4	16
$h=2^{-4}$	0.012	0.130	0.200	$h=2^{-4}$	0.003	0.064	0.156	$h=2^{-4}$	0.004	0.012	0.083
$h=2^{-5}$	0.021	0.129	0.177	$h=2^{-5}$	0.013	0.132	0.214	$h=2^{-5}$	0.014	0.091	0.168
$h=2^{-6}$	0.021	0.131	0.133	$h=2^{-6}$	0.015	0.143	0.187	$h=2^{-6}$	0.031	0.140	0.223

Table 2 Spectral radius ρ for the second benchmark for different values of h and p for a different number of patches

	# patches				# patches				# patches		
$p=2$	1	4	16	$p=3$	1	4	16	$p=4$	1	4	16
$h=2^{-4}$	0.014	0.049	0.141	$h=2^{-4}$	0.003	0.013	0.073	$h=2^{-4}$	0.003	0.007	1.312
$h=2^{-5}$	0.039	0.093	0.155	$h=2^{-5}$	0.020	0.073	0.155	$h=2^{-5}$	0.029	0.035	0.090
$h=2^{-6}$	0.057	0.103	0.116	$h=2^{-6}$	0.024	0.124	0.169	$h=2^{-6}$	0.023	0.114	0.174

Table 3 Spectral radius ρ for the third benchmark for different values of h and p for a different number of patches

	# patches				# patches				# patches		
$p=2$	1	4	16	$p=3$	1	4	16	$p=4$	1	4	16
$h=2^{-4}$	0.004	0.038	0.080	$h=2^{-4}$	0.001	0.002	0.013	$h=2^{-4}$	$2.6\cdot 10^{-5}$	$6.5\cdot 10^{-5}$	0.003
$h=2^{-5}$	0.012	0.082	0.129	$h=2^{-5}$	0.007	0.035	0.080	$h=2^{-5}$	0.002	0.005	0.020
$h=2^{-6}$	0.016	0.089	0.127	$h=2^{-6}$	0.010	0.091	0.159	$h=2^{-6}$	0.005	0.059	0.118

5 Numerical Results

The proposed p-multigrid method is applied in this section as a stand-alone solver and as a preconditioner within a Biconjugate gradient stabilized (BiCGSTAB) method. Results are obtained using different numbers of patches. Furthermore, results are compared when using ILUT as a solver. Finally, different coarsening strategies (i.e. coarsening in h, p or both) are compared with respect to the number of iterations and computational time.

The initial guess $\mathbf{u}_{h,p}^{(0)}$ is chosen randomly for all experiments, where each entry is sampled from a uniform distribution on the interval $[-1, 1]$ using the same seed. The method is considered to be converged if

$$\frac{||\mathbf{r}_{h,p}^{(n)}||}{||\mathbf{r}_{h,p}^{(0)}||} < 10^{-8}, \qquad (24)$$

where $\mathbf{r}_{h,p}^{(n)}$ denotes the residual after iteration n. The solution of the residual equation is obtained at level $p = 1$ by means of a Conjugate Gradient solver with a relatively high tolerance ($\epsilon = 10^{-4}$). The same number of pre- and postsmoothing

Table 4 Number of V-cycles needed for achieving convergence for the first benchmark with p-multigrid, $\nu = 2$

$p = 2$	# patches			$p = 3$	# patches			$p = 4$	# patches		
	1	4	16		1	4	16		1	4	16
$h = 2^{-4}$	3	5	7	$h = 2^{-4}$	2	3	5	$h = 2^{-4}$	2	2	4
$h = 2^{-5}$	3	6	7	$h = 2^{-5}$	3	5	7	$h = 2^{-5}$	2	4	5
$h = 2^{-6}$	3	5	6	$h = 2^{-6}$	3	6	7	$h = 2^{-6}$	3	5	7
$h = 2^{-7}$	3	5	5	$h = 2^{-7}$	3	5	5	$h = 2^{-7}$	2	5	6

Table 5 Number of V-cycles needed for reaching convergence for the first benchmark with p-multigrid, $\nu = 4$

$p = 2$	# patches			$p = 3$	# patches			$p = 4$	# patches		
	1	4	16		1	4	16		1	4	16
$h = 2^{-4}$	2	4	5	$h = 2^{-4}$	1	2	3	$h = 2^{-4}$	1	2	2
$h = 2^{-5}$	3	5	6	$h = 2^{-5}$	2	4	5	$h = 2^{-5}$	2	3	3
$h = 2^{-6}$	2	5	6	$h = 2^{-6}$	2	5	6	$h = 2^{-6}$	2	4	5
$h = 2^{-7}$	2	4	4	$h = 2^{-7}$	2	4	5	$h = 2^{-7}$	2	4	5

steps are applied for all experiments ($\nu = \nu_1 = \nu_2$) and boundary conditions are imposed using Nitsche's method [12].

The number of V-cycles needed to reach convergence for different values of h and p for a different number of patches for the first benchmark are shown in Table 4. Here, the number of smoothing steps at each level equals 2. Results have been obtained considering 1, 4 and 16 patches, where the multipatch is based on splitting a single patch uniformly in both directions. In general, the p-multigrid method shows convergence rates which are essentially independent of both h and p. However, an increase of the number of patches leads to an increase in the number of V-cycles needed to achieve convergence. Note that this increase is relatively low, especially for smaller values of h.

Table 5 shows the number of V-cycles needed when the number of smoothing steps is doubled. Hence, $\nu = 4$ for all numerical experiments. Doubling the number of smoothing steps at each level, slightly decreases the number of V-cycles. However, since the number of V-cycles is already relatively low, the reduction is limited.

The number of V-cycles needed to achieve convergence for the second benchmark are presented in Table 6. Compared to the first benchmark, convergence is obtained in the same or even a lower number of V-cycles. Furthermore, only a small increase of the number of V-cycles needed is observed when the number of patches increases. For one configuration, however, the p-multigrid method diverges, as expected based on the spectral analysis; see Table 2.

Table 7 presents the number of V-cycles needed to achieve convergence for the third benchmark. Again, convergence is reached in a relatively low number of V-cycles. Furthermore, no instabilities are observed for the considered configurations.

Table 6 Number of V-cycles needed to achieve convergence for the second benchmark with p-multigrid, $\nu = 2$

$p=2$	# patches			$p=3$	# patches			$p=4$	# patches		
	1	4	16		1	4	16		1	4	16
$h=2^{-4}$	2	3	5	$h=2^{-4}$	2	2	3	$h=2^{-4}$	1	2	–
$h=2^{-5}$	3	4	5	$h=2^{-5}$	2	3	4	$h=2^{-5}$	2	2	3
$h=2^{-6}$	3	4	5	$h=2^{-6}$	2	4	5	$h=2^{-6}$	2	4	5
$h=2^{-7}$	3	4	4	$h=2^{-7}$	2	4	5	$h=2^{-7}$	2	4	5

Table 7 Number of V-cycles needed to achieve convergence for the third benchmark with p-multigrid, $\nu = 2$

$p=2$	# patches			$p=3$	# patches			$p=4$	# patches		
	1	4	16		1	4	16		1	4	16
$h=2^{-4}$	3	4	5	$h=2^{-4}$	2	2	3	$h=2^{-4}$	2	2	3
$h=2^{-5}$	3	5	6	$h=2^{-5}$	2	4	5	$h=2^{-5}$	2	3	3
$h=2^{-6}$	3	4	5	$h=2^{-6}$	2	5	6	$h=2^{-6}$	2	4	5
$h=2^{-7}$	3	4	5	$h=2^{-7}$	2	4	5	$h=2^{-7}$	2	4	5

Table 8 Number of iterations needed to reach convergence for the first benchmark with preconditioned BiCGSTAB, $\nu = 2$

$p=2$	# patches			$p=3$	# patches			$p=4$	# patches		
	1	4	16		1	4	16		1	4	16
$h=2^{-4}$	2	2	2	$h=2^{-4}$	1	2	2	$h=2^{-4}$	1	1	2
$h=2^{-5}$	2	2	3	$h=2^{-5}$	1	2	2	$h=2^{-5}$	1	2	2
$h=2^{-6}$	2	2	3	$h=2^{-6}$	2	2	3	$h=2^{-6}$	1	2	2
$h=2^{-7}$	2	2	2	$h=2^{-7}$	1	2	2	$h=2^{-7}$	1	2	2

The increase in the number of V-cycles needed to reach convergence when the number of patches is increased is only mild.

Alternatively, the p-multigrid method is applied as a preconditioner within a Biconjugate gradient stabilized (BiCGSTAB) method. At the preconditioning phase of every iteration of the BiCGSTAB method, a single V-cycle is applied. Again, a tolerance of $\epsilon = 10^{-8}$ is adopted as a stopping criterion for the BiCGSTAB solver.

Results obtained for the first benchmark are presented in Table 8. Compared to the use of p-multigrid as a stand-alone solver, the number of iterations needed on a single patch geometry is smaller for all configurations. In case of a multipatch geometry, however, the number of iterations needed reduces even more when a BiCGSTAB method is adopted. Hence, the difference in BiCGSTAB iterations for single patch and multipatch geometries becomes even smaller.

For the second and third benchmark, results are presented in Tables 9 and 10, respectively. For the single patch geometry, the number of iterations with the BiCGSTAB method is again smaller compared to the number of V-cycles for the p-multigrid method for almost all configurations. A slightly larger reduction in the number of iterations can be observed for some numerical experiments in case of a

Table 9 Number of iterations needed for reaching convergence for the second benchmark with preconditioned BiCGSTAB, $\nu = 2$

	# patches				# patches				# patches		
$p=2$	1	4	16	$p=3$	1	4	16	$p=4$	1	4	16
$h=2^{-4}$	1	2	2	$h=2^{-4}$	1	1	2	$h=2^{-4}$	1	1	3
$h=2^{-5}$	2	2	2	$h=2^{-5}$	1	1	2	$h=2^{-5}$	1	1	2
$h=2^{-6}$	2	2	2	$h=2^{-6}$	1	2	2	$h=2^{-6}$	1	2	2
$h=2^{-7}$	2	2	2	$h=2^{-7}$	1	2	2	$h=2^{-7}$	1	2	2

Table 10 Number of iterations needed for reaching convergence for the third benchmark with preconditioned BiCGSTAB, $\nu = 2$

	# patches				# patches				# patches		
$p=2$	1	4	16	$p=3$	1	4	16	$p=4$	1	4	16
$h=2^{-4}$	1	2	2	$h=2^{-4}$	1	1	2	$h=2^{-4}$	1	1	1
$h=2^{-5}$	2	2	3	$h=2^{-5}$	1	2	2	$h=2^{-5}$	1	1	2
$h=2^{-6}$	2	2	3	$h=2^{-6}$	1	2	3	$h=2^{-6}$	1	2	2
$h=2^{-7}$	2	2	2	$h=2^{-7}$	1	2	2	$h=2^{-7}$	1	2	2

multipatch geometry. Note that BiCGSTAB restores stability for the setting in which the p-multigrid algorithm separately is unstable; see Table 2.

As discussed in Sect. 3, different cycle types can be adopted. The use of a W-cycle instead of the V-cycle leads to the same number of cycles needed for all numerical experiments. Considering the higher computational costs for a single W-cycle, a V-cycle is adopted throughout the rest of this paper.

5.1 ILUT as a Solver

The obtained results are compared to using ILUT as a stand-alone solver. In this way, the effectiveness of the coarse grid correction within the p-multigrid method can be investigated. Table 11 shows the number of iterations needed to achieve convergence with ILUT as a solver. For all numerical experiments, the number of iterations needed with ILUT is significantly higher compared to the number of V-cycles needed with p-multigrid (see Table 4 for comparison). Furthermore, the number of iterations needed with ILUT as a solver becomes h-dependent, leading to a high number of iterations on finer meshes. As shown in Tables 12 and 13, the same observations can be made for the second and third benchmark, respectively. These results indicate that the coarse grid correction is necessary to obtain a low number of iterations until convergence in reached.

Table 11 Number of iterations needed for achieving convergence for the first benchmark with ILUT as solver

$p=2$	# patches			$p=3$	# patches			$p=4$	# patches		
	1	4	16		1	4	16		1	4	16
$h=2^{-4}$	25	40	55	$h=2^{-4}$	12	18	29	$h=2^{-4}$	7	10	18
$h=2^{-5}$	96	125	148	$h=2^{-5}$	50	53	67	$h=2^{-5}$	22	27	35
$h=2^{-6}$	352	397	437	$h=2^{-6}$	171	182	199	$h=2^{-6}$	80	86	106
$h=2^{-7}$	1280	1356	1440	$h=2^{-7}$	609	623	664	$h=2^{-7}$	288	307	324

Table 12 Number of iterations needed to obtain convergence for the second benchmark with ILUT as solver

$p=2$	# patches			$p=3$	# patches			$p=4$	# patches		
	1	4	16		1	4	16		1	4	16
$h=2^{-4}$	14	17	28	$h=2^{-4}$	6	9	15	$h=2^{-4}$	4	6	--
$h=2^{-5}$	56	55	63	$h=2^{-5}$	23	21	30	$h=2^{-5}$	12	13	16
$h=2^{-6}$	194	219	217	$h=2^{-6}$	76	84	83	$h=2^{-6}$	37	39	41
$h=2^{-7}$	716	710	700	$h=2^{-7}$	251	276	301	$h=2^{-7}$	131	138	137

Table 13 Number of iterations needed to obtain convergence for the third benchmark with ILUT as solver

$p=2$	# patches			$p=3$	# patches			$p=4$	# patches		
	1	4	16		1	4	16		1	4	16
$h=2^{-4}$	33	26	32	$h=2^{-4}$	13	10	16	$h=2^{-4}$	8	7	12
$h=2^{-5}$	126	90	88	$h=2^{-5}$	45	25	33	$h=2^{-5}$	23	14	17
$h=2^{-6}$	469	290	288	$h=2^{-6}$	168	88	100	$h=2^{-6}$	80	40	47
$h=2^{-7}$	1667	1046	1050	$h=2^{-7}$	596	320	332	$h=2^{-7}$	283	150	146

5.2 Comparison h- and hp-Multigrid

In the previous subsection, it was shown that a coarse grid correction is necessary to obtain an efficient solution method. To determine the quality of the coarse grid correction with p-multigrid in more detail, results are compared with h- and hp-multigrid methods. In these methods, only the way in which the hierarchy is constructed differs. For the h-multigrid method, coarsening in h is applied, while for the hp-multigrid method coarsening in h and p is applied simultaneously. All other components (i.e. smoothing, prolongation and restriction) are identical. It should be noted that, since coarsening in h leads to a nested hierarchy of discretizations, a canonical prolongation/restriction operator could be defined for the h-multigrid method. These transfer operators are, however, not taken into account in this paper. Results obtained for the benchmarks with the different coarsening strategies on a multipatch geometry for different values of h and p are shown in Tables 14, 15, and 16, respectively.

Table 14 Comparison of p-multigrid with h- and hp-multigrid for the first benchmark on 4 patches, $\nu = 2$

$p = 2$	p	h	hp	$p = 3$	p	h	hp	$p = 4$	p	h	hp
$h = 2^{-3}$	5	8	9	$h = 2^{-3}$	3	4	4	$h = 2^{-3}$	2	3	3
$h = 2^{-4}$	6	18	20	$h = 2^{-4}$	5	10	11	$h = 2^{-4}$	4	6	6
$h = 2^{-5}$	5	28	31	$h = 2^{-5}$	6	25	31	$h = 2^{-5}$	5	14	16
$h = 2^{-6}$	5	32	35	$h = 2^{-6}$	5	56	70	$h = 2^{-6}$	5	36	48

Table 15 Comparison of p-multigrid with h- and hp-multigrid for the second benchmark on 4 patches, $\nu = 2$

$p = 2$	p	h	hp	$p = 3$	p	h	hp	$p = 4$	p	h	hp
$h = 2^{-3}$	3	4	4	$h = 2^{-3}$	2	2	3	$h = 2^{-3}$	2	2	2
$h = 2^{-4}$	4	10	11	$h = 2^{-4}$	3	5	5	$h = 2^{-4}$	2	3	3
$h = 2^{-5}$	4	20	22	$h = 2^{-5}$	4	13	16	$h = 2^{-5}$	4	8	8
$h = 2^{-6}$	4	26	27	$h = 2^{-6}$	4	31	39	$h = 2^{-6}$	4	19	23

Table 16 Comparison of p-multigrid with h- and hp-multigrid for the third benchmark on 4 patches, $\nu = 2$

$p = 2$	p	h	hp	$p = 3$	p	h	hp	$p = 4$	p	h	hp
$h = 2^{-3}$	4	6	6	$h = 2^{-3}$	2	3	3	$h = 2^{-3}$	2	2	2
$h = 2^{-4}$	5	14	16	$h = 2^{-4}$	4	6	6	$h = 2^{-4}$	3	3	3
$h = 2^{-5}$	4	23	25	$h = 2^{-5}$	5	14	17	$h = 2^{-5}$	4	8	8
$h = 2^{-6}$	4	28	30	$h = 2^{-6}$	4	34	46	$h = 2^{-6}$	4	20	25

For all benchmarks, the number of V-cycles needed with p-multigrid is significantly lower for all configurations compared to h- and hp-multigrid. Furthermore, the difference in the number of V-cycles increases when the knot span size is halved. In general, coarsening in h is more efficient compared to coarsening both in h and p. Results indicate that the coarsening in p leads to the most effective coarse grid correction, resulting in the lowest number of V-cycles.

Besides the number of iterations, CPU times have been compared for the different multigrid methods. This comparison is in particular interesting, since the coarse grid correction is more expensive for p-multigrid methods compared to h- and hp-multigrid approaches. A serial implementation is considered on a Intel(R) Xeon(R) E5-2687W CPU (3.10 GHz). Tables 17, 18, and 19 present the computational times (in seconds) for all benchmarks obtained with the different coarsening strategies.

Table 17 Computational times (in seconds) with p, h- and hp-multigrid for the first benchmark on 4 patches, $\nu = 2$

$p = 2$	p	h	hp	$p = 3$	p	h	hp	$p = 4$	p	h	hp
$h = 2^{-3}$	1.4	1.0	1.2	$h = 2^{-3}$	1.4	1.4	1.4	$h = 2^{-3}$	2.0	1.8	1.9
$h = 2^{-4}$	4.2	4.9	4.5	$h = 2^{-4}$	5.1	5.8	5.8	$h = 2^{-4}$	7.4	7.9	6.8
$h = 2^{-5}$	13.1	22.1	21.9	$h = 2^{-5}$	21.9	38.7	40.7	$h = 2^{-5}$	31.6	48.0	44.7
$h = 2^{-6}$	62.0	126.3	127.7	$h = 2^{-6}$	105.1	384.2	419.5	$h = 2^{-6}$	169.1	508.8	542.4

Table 18 Computational times (in seconds) with p-, h- and hp-multigrid for the second benchmark on 4 patches, $v = 2$

$p = 2$	p	h	hp	$p = 3$	p	h	hp	$p = 4$	p	h	hp
$h = 2^{-3}$	1.2	1.0	1.1	$h = 2^{-3}$	1.6	1.5	1.7	$h = 2^{-3}$	2.2	1.8	1.9
$h = 2^{-4}$	4.4	4.9	4.8	$h = 2^{-4}$	5.7	5.9	5.3	$h = 2^{-4}$	7.3	7.7	6.9
$h = 2^{-5}$	14.4	22.1	23.0	$h = 2^{-5}$	22.1	33.7	33.5	$h = 2^{-5}$	35.5	42.4	36.8
$h = 2^{-6}$	64.2	142.3	136.7	$h = 2^{-6}$	105.6	306.3	320.8	$h = 2^{-6}$	175.8	382.6	376.6

Table 19 Computational times (in seconds) with p-, h- and hp-multigrid for the third benchmark on 4 patches, $v = 2$

$p = 2$	p	h	hp	$p = 3$	p	h	hp	$p = 4$	p	h	hp
$h = 2^{-3}$	1.1	1.1	1.0	$h = 2^{-3}$	1.3	1.3	1.6	$h = 2^{-3}$	2.0	1.8	1.7
$h = 2^{-4}$	4.2	4.8	3.9	$h = 2^{-4}$	5.4	5.2	5.0	$h = 2^{-4}$	7.1	6.4	6.1
$h = 2^{-5}$	13.1	20.0	20.1	$h = 2^{-5}$	20.5	27.3	27.9	$h = 2^{-5}$	29.1	32.9	29.9
$h = 2^{-6}$	59.8	120.1	119.7	$h = 2^{-6}$	94.3	257.5	299.1	$h = 2^{-6}$	149.0	307.6	308.8

On coarser grids, the computational times for all multigrid methods are comparable. For smaller values of h, however, the computational time needed with p-multigrid is significantly smaller compared to h- and hp-multigrid due to the considerable h-dependency of the latter two approaches. Furthermore, the computational time needed with p-multigrid scales (almost) linearly with the number of degrees of freedom. This holds for all benchmarks and all values of p considered in this study.

6 Conclusion

In this paper, we have extended our p-multigrid solver for IgA discretizations using a Incomplete LU factorization [19] to multipatch geometries. An analysis of the spectrum of the iteration matrix shows that this p-multigrid method can be applied on multipatch geometries, with convergence rates essentially independent of the knot span size h and approximation order p. Only a mild dependence of the convergence rate on the number of patches is observed. Numerical results, obtained for Poisson's equation on the unit square, the quarter annulus and an L-shaped domain, confirm this analysis. Furthermore, results show the necessity of the coarse grid correction within the p-multigrid method. Finally, different coarsening strategies have been compared, indicating that coarsening in p is most effective compared to coarsening in h or h and p simultaneously. Future research should focus on the application of the p-multigrid method on partial differential equations of higher-order, for example the biharmonic equation. Furthermore, the use of p-multigrid in a HPC framework can be investigated, in which block ILUT can be applied efficiently as a smoother on each multipatch separately.

Acknowledgments The authors would like to thank both Prof. Kees Oosterlee from TU Delft and Prof. Dominik Göddeke from the University of Stuttgart for fruitfull discussions with respect to p-multigrid methods.

References

1. Brenner, S.C. and Scott, L.R.: The mathematical theory of finite element methods, Texts Applied Mathematics, 15, Springer, New York (1994).
2. Briggs, W.L., Henson, V. E. and McCormick, S.F.: A Multigrid Tutorial 2nd edition, SIAM, Philadelphia, (2000)
3. Donatelli, M., Garoni, C., Manni, C., Capizzano, S. and Speleers,H.: Symbol-based multigrid methods for Galerkin B-spline isogeometric analysis. SIAM Journal on Numerical Analysis, **55**(1), 31–62 (2017)
4. Fidkowski, K.J., Oliver, T.A., Lu. J and Darmofal, D.L.: p-Multigrid solution of high-order discontinuous Galerkin discretizations of the compressible Navier-Stokes equations. Journal of Computational Physics, **207**(1), 92–113 (2005)
5. Gahalaut, K.P.S., Kraus, J.K. and Tomar, S.K.: Multigrid methods for isogeometric discretizations. Computer Methods in Applied Mechanics and Engineering, **253**, 413–425 (2013)
6. Guennebaud, G., Benoît, J. et al.: Eigen v3, http://eigen.tuxfamily.org, (2010)
7. Helenbrook, B., Mavriplis, D. and Atkins, H.: Analysis of p-Multigrid for Continuous and Discontinuous Finite Element Discretizations. 16th AIAA Computational Fluid Dynamics Conference, Fluid Dynamics and Co-located Conferences
8. Hofreither, C., Takacs, S. and Zulehner, W.: A robust multigrid method for Isogeometric Analysis in two dimensions using boundary correction. Computer Methods in Applied Mechanics and Engineering, **316**, 22–42 (2017)
9. Hughes, T.J.R., Cottrell J.A. and Bazilevs, Y.: Isogeometric analysis: CAD, finite elements, NURBS, exact geometry and mesh refinement. Computer Methods in Applied Mechanics and Engineering. **194**, 4135–4195 (2005)
10. Luo, H., Baum, J.D. and Löhner, R.: A p-multigrid discontinuous Galerkin method for the Euler equations on unstructured grids. Journal of Computational Physics, **211**(2), 767–783 (2006)
11. Luo, H., Baum, J.D. and Löhner, R.: Fast p-Multigrid Discontinuous Galerkin Method for Compressible Flows at All Speeds. AIAA Journal, **46**(3), 635–652 (2008)
12. Nitsche, J.: Über ein Variationsprinzip zur Lösung von Dirichlet-Problemen bei Verwendung von Teilräumen die keinen Randbedingungen unterworfen sind. Abhandlungen aus dem mathematischen Seminar der Universität Hamburg, **36**(1), 9–15 (1971)
13. de la Riva, A., Rodrigo, C. and Gaspar, F.: An efficient multigrid solver for isogeometric analysis. arXiv:1806.05848v1, 2018
14. van Slingerland, P. and Vuik, C.: Fast linear solver for diffusion problems with applications to pressure computation in layered domains. Computational Geosciences, **18**(3–4), 343–356 (2014)
15. Saad, Y.: ILUT: A dual threshold incomplete LU factorization. Numerical Linear Algebra with Applications, **1**(4), 387–402 (1994)
16. Saad, Y.: SPARSKIT: A basic tool kit for sparse matrix computations. (1994)
17. Sampath, R.S. and Biros, G.: A parallel geometric multigrid method for finite elements on octree meshes, SIAM Journal on Scientific Computing, **32**(3), 1361–1392 (2010)
18. Tielen, R., Möller, M. and Vuik, C.: Efficient multigrid based solvers for Isogeometric Analysis. Proceedings of the 6th European Conference on Computational Mechanics and the 7th European Conference on Computational Fluid Dynamics, Glasgow, UK, 2018
19. Tielen, R., Möller, M., Göddeke, D. and Vuik, C.: p-multigrid methods and their comparison to h-multigrid methods within Isogeometric Analysis. Computer Methods in Applied Mechanics and Engineering, **372** (2020)
20. Trottenberg, U., Oosterlee, C. and Schüller, A.: Multigrid, Academic Press (2001).

The Use of Dual B-Spline Representations for the Double de Rham Complex of Discrete Differential Forms

Yi Zhang, Varun Jain, Artur Palha, and Marc Gerritsma

Abstract In $\mathbb{R}^n$, let $\Lambda^k(\Omega)$ represent the space of smooth differential k-forms in Ω. The de Rham complex consists of a sequence of spaces, $\Lambda^k(\Omega)$, $k = 0, 1 \ldots, n$, connected by the exterior derivative, d : $\Lambda^k(\Omega) \to \Lambda^{k+1}(\Omega)$. Appropriately chosen B-spline spaces together with their associated dual B-spline spaces form a discrete double de Rham complex. In practical applications, this discrete double de Rham complex leads to very sparse systems. In this paper, this construction will be explained and illustrated by means of a non-trivial three-dimensional example.

1 Introduction

Given a bounded domain Ω in $\mathbb{R}^n$, spaces of smooth differential forms, $\Lambda^k(\Omega)$, $k = 0, 1, \cdots, n$, and the exterior derivative d compose an exact sequence,

$$\mathbb{R} \hookrightarrow \Lambda^0(\Omega) \xrightarrow{\mathrm{d}} \Lambda^1(\Omega) \xrightarrow{\mathrm{d}} \cdots \xrightarrow{\mathrm{d}} \Lambda^n(\Omega) \longrightarrow 0 \,,$$

called the de Rham complex [1, 3, 10, 13, 16].

Consider finite dimensional Hilbert spaces $W^k \subset \Lambda^k(\Omega)$ and the linear operator d: $W^k \to W^{k+1}$ such that the range of d on W^k, $\mathfrak{B}^{k+1}$, is contained in the null space of d on W^{k+1}, $\mathfrak{Z}^{k+1}$, i.e. $\mathfrak{B}^{k+1} \subset \mathfrak{Z}^{k+1}$ [1]. The space of harmonic forms is defined as $\mathfrak{H}^k := \mathfrak{B}^{k,\perp} \cap \mathfrak{Z}^k$. With these subspaces we have the Hodge decomposition,

$$W^k = \mathfrak{B}^k \oplus \mathfrak{H}^k \oplus \mathfrak{Z}^{k,\perp} \,.$$

Y. Zhang (✉) · V. Jain · A. Palha · M. Gerritsma
Delft University of Technology, Delft, The Netherlands
e-mail: y.zhang-14@tudelft.nl; v.jain@tudelft.nl; A.Palha@tudelft.nl; m.i.gerritsma@tudelft.nl

© Springer Nature Switzerland AG 2021
H. van Brummelen et al. (eds.), *Isogeometric Analysis and Applications 2018*,
Lecture Notes in Computational Science and Engineering 133,
https://doi.org/10.1007/978-3-030-49836-8_11

Since we are in finite dimensional function spaces, the operator d is trivially bounded and all the spaces W^k are complete. A de Rham complex then can be depicted as

$$\mathbb{R} \hookrightarrow W^0 \xrightarrow{d} W^1 \xrightarrow{d} \cdots \xrightarrow{d} W^n \longrightarrow 0,$$

where the real numbers on the left of the sequence denote the null space of d on W^0.

With each function space W^k, we will associate a dual function space $\widetilde{W}^k$ of linear functionals acting on the elements of W^k. The linear operator d acting on the spaces W^k induces an adjoint operator $\widetilde{d}$ acting on the dual spaces $\widetilde{W}^k$. Let $\widetilde{\alpha}^{k+1} \in \widetilde{W}^{k+1}$, then $\widetilde{d}$ is defined by Kreyszig [14],

$$\left(\widetilde{d}\,\widetilde{\alpha}^{k+1}\right)(\omega) := \widetilde{\alpha}^{k+1}(d\omega), \quad \forall \omega \in W^k. \tag{1}$$

So $\widetilde{d} : \widetilde{W}^{k+1} \longrightarrow \widetilde{W}^k$. The dual spaces and the adjoint operator form an adjoint de Rham complex:

$$0 \longleftarrow \widetilde{W}^0 \xleftarrow{\widetilde{d}} \widetilde{W}^1 \xleftarrow{\widetilde{d}} \cdots \xleftarrow{\widetilde{d}} \widetilde{W}^n \hookleftarrow \mathbb{R}.$$

Since we are now working with Hilbert spaces, Riesz' representation theorem states that for every linear functional $\widetilde{\alpha}^k \in \widetilde{W}^k$, there exists a unique element $\omega_\alpha \in W^k$, such that

$$\widetilde{\alpha}^k(\omega) = (\omega_\alpha, \omega)_{W^k}, \quad \forall \omega \in W^k, \tag{2}$$

where $(\cdot, \cdot)_{W^k}$ denotes an inner product on W^k. If $\boldsymbol{g}^k$ denotes the metric tensor on W^k, then this relation can be written as $\widetilde{\alpha}^k = \boldsymbol{g}^k \omega_\alpha$. This connection between the primal and the dual spaces leads to the double de Rham complex shown in Fig. 1.

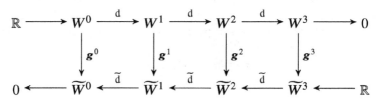

Fig. 1 The double de Rham complex

In the current paper, the finite dimensional spaces $W^k \subset \Lambda^k(\Omega)$ will be spanned by B-splines, see [5]. The construction of their dual representations will also be introduced. These B-spline spaces and their dual representations form a particular instance of the double de Rham complex given in Fig. 1. The use of this particular instance leads to very sparse discrete systems which can preserve the exterior derivative d.

The layout of this paper is as follows: In Sect. 2, we introduce the construction of B-spline spaces and their dual representations. They are used to solve the Poisson problem in a mixed formulation in Sect. 3. Conclusions are drawn in Sect. 4.

2 B-Spline Spaces

2.1 Primal B-Spline Spaces

Let p and N, $1 \leq p \leq N$, be two positive integers. We consider a p-open knot vector, $\Xi := \{\xi_0, \xi_1, \cdots, \xi_{N+p+1}\}$, where

$$\xi_0 = \xi_1 = \cdots = \xi_p < \xi_{p+1} < \cdots < \xi_{N+1} = \xi_{N+2} = \cdots = \xi_{N+p+1}.$$

With Ξ, $(N+1)$ B-spline basis functions of degree p, $B_\Xi^{i,p}$, $i = 0, 1, \cdots, N$, can be derived by the well-known Cox-de Boor recursion [6, 7]: For $p = 0$,

$$B_\Xi^{i,0}(\xi) = \begin{cases} 1 & \text{if } \xi_i \leq \xi < \xi_{i+1} \\ 0 & \text{otherwise} \end{cases},$$

and for $p \geq 1$,

$$B_\Xi^{i,p}(\xi) = \frac{\xi - \xi_i}{\xi_{i+p} - \xi_i} B_\Xi^{i,p-1}(\xi) + \frac{\xi_{i+p+1} - \xi}{\xi_{i+p+1} - \xi_{i+1}} B_\Xi^{i+1,p-1}(\xi).$$

These $(N+1)$ linearly independent B-spline basis functions span a finite dimensional B-spline space, $\mathcal{S}_\Xi^p$,

$$\mathcal{S}_\Xi^p := \text{span}\left\{B_\Xi^{i,p}\right\}.$$

The Greville points of the p-open knot vector Ξ are obtained by

$$\tau_i = \frac{1}{p}\sum_{j=1}^{p} \xi_{i+j}, \ i = 0, 1, \cdots, N.$$

If $\varphi^{(0)}(\xi)$ is a continuous 0-form. The projection of $\varphi^{(0)}$ in $\mathcal{S}_\Xi^p$, $\varphi_h^{(0)}$, can be expressed as

$$\varphi_h^{(0)}(\xi) = \sum_{i=0}^{N} \varphi_i B_\Xi^{i,p}(\xi),$$

whose coefficients, φ_i, can be computed by

$$(\varphi_0, \varphi_1, \cdots, \varphi_N)^\mathsf{T} = \mathbb{B}^{-1} \left(\varphi^{(0)}(\tau_0), \varphi^{(0)}(\tau_1), \cdots, \varphi^{(0)}(\tau_N)\right)^\mathsf{T},$$

where $\mathbb{B}$ is an $(N+1)$ by $(N+1)$ matrix and $\mathbb{B}_{ij} = B_\Xi^{j,p}(\tau_i)$.

We use Ξ' to denote the $(p-1)$-open knot vector, $\Xi' := \{\xi_1, \xi_2, \cdots, \xi_{N+p}\}$. From Ξ', the Cox-de Boor recursion gives N B-spline basis functions of degree $(p-1)$, $B_{\Xi'}^{j,p-1}$, $j = 1, 2, \cdots, N$. We can further construct the Curry-Schoenberg B-spline basis functions, [4, 5, 11], by scaling $B_{\Xi'}^{j,p-1}$ as

$$M_{\Xi'}^{j,p-1} := \frac{p}{\xi_{p+j} - \xi_j} B_{\Xi'}^{j,p-1}, \quad j = 1, 2, \cdots, N.$$

These N B-spline basis functions form a basis of a B-spline space, denoted by $\mathcal{S}_{\Xi'}^{p-1}$,

$$\mathcal{S}_{\Xi'}^{p-1} := \mathrm{span}\left\{M_{\Xi'}^{j,p-1}\right\}.$$

Let a continuous 1-form, $u^{(1)}(\xi)$, be expanded in $\mathcal{S}_{\Xi'}^{p-1}$,

$$u_h^{(1)}(\xi) = \sum_{j=1}^{N} u_j M_{\Xi'}^{j,p-1}(\xi) \mathrm{d}\xi,$$

whose coefficients, u_j, can be computed by

$$(u_1, u_2, \cdots, u_N)^\mathsf{T} = \mathbb{M}^{-1} \left(\int_{\tau_0}^{\tau_1} u^{(1)}(\xi), \int_{\tau_1}^{\tau_2} u^{(1)}(\xi), \cdots, \int_{\tau_{N-1}}^{\tau_N} u^{(1)}(\xi)\right)^\mathsf{T},$$

where $\mathbb{M}$ is an N by N matrix and $\mathbb{M}_{ij} = \int_{\tau_{i-1}}^{\tau_i} M_{\Xi'}^{j,p-1}(\xi)\mathrm{d}\xi$. For the condition of $\mathbb{B}$ and $\mathbb{M}$ being non-singular, we refer to [2, 7].

Furthermore, if $\varphi^{(0)}$ and $u^{(1)}$ satisfy $u^{(1)} = d\varphi^{(0)}$, their discrete formulations, $\varphi_h^{(0)}$ and $u_h^{(1)}$, will satisfy

$$u_h^{(1)} = \sum_{j=1}^{N} u_j M_{\Xi'}^{j,p-1}(\xi)d\xi = \sum_{j=1}^{N} \left(\varphi_j - \varphi_{j-1}\right) M_{\Xi'}^{j,p-1}(\xi)d\xi = d\varphi_h^{(0)},$$

which implies that the range of the exterior derivative d on $\mathcal{S}_\Xi^p$ is in $\mathcal{S}_{\Xi'}^{p-1}$. In other words, $\mathcal{S}_\Xi^p$ and $\mathcal{S}_{\Xi'}^{p-1}$ form a discrete de Rham complex in $\mathbb{R}^1$,

$$\mathbb{R} \hookrightarrow \mathcal{S}_\Xi^p \xrightarrow{d} \mathcal{S}_{\Xi'}^{p-1} \longrightarrow 0.$$

Furthermore, let $\boldsymbol{\varphi}$ and $\boldsymbol{u}$ be the vectors of the expansion coefficients of $\varphi_h^{(0)}$ and $u_h^{(1)}$. We have

$$\boldsymbol{u} = \begin{pmatrix} -1 & 1 & 0 & \cdots & 0 \\ 0 & -1 & 1 & \cdots & 0 \\ \vdots & \vdots & \vdots & \ddots & \vdots \\ 0 & \cdots & 0 & -1 & 1 \end{pmatrix} \boldsymbol{\varphi} = \mathbb{E}^{1,0} \boldsymbol{\varphi},$$

where $\mathbb{E}^{1,0}$, called the incidence matrix, is very sparse and only depends on the topology of the mesh. Notice that the relation between $\varphi_h^{(0)}$ and its derivative $u_h^{(1)}$ is exact, which means that the incidence matrix is an exact discrete counterpart of the exterior derivative d.

The B-spline spaces for higher dimensional spaces are constructed using the tensor product. For example, in $\mathbb{R}^3$ with coordinate system $\{\xi^1, \xi^2, \xi^3\}$, let Ξ_1, Ξ_2 and Ξ_3 be p_1-, p_2- and p_3-open knot vectors along ξ^1, ξ^2 and ξ^3 respectively. We define

$$\mathcal{S}^0_{\Xi_1,\Xi_2,\Xi_3} := \mathcal{S}^{p_1,p_2,p_3}_{\Xi_1,\Xi_2,\Xi_3},$$

$$\mathcal{S}^1_{\Xi_1,\Xi_2,\Xi_3} := \mathcal{S}^{p_1-1,p_2,p_3}_{\Xi'_1,\Xi_2,\Xi_3} \times \mathcal{S}^{p_1,p_2-1,p_3}_{\Xi_1,\Xi'_2,\Xi_3} \times \mathcal{S}^{p_1,p_2,p_3-1}_{\Xi_1,\Xi_2,\Xi'_3},$$

$$\mathcal{S}^2_{\Xi_1,\Xi_2,\Xi_3} := \mathcal{S}^{p_1,p_2-1,p_3-1}_{\Xi_1,\Xi'_2,\Xi'_3} \times \mathcal{S}^{p_1-1,p_2,p_3-1}_{\Xi'_1,\Xi_2,\Xi'_3} \times \mathcal{S}^{p_1-1,p_2-1,p_3}_{\Xi'_1,\Xi'_2,\Xi_3},$$

$$\mathcal{S}^3_{\Xi_1,\Xi_2,\Xi_3} := \mathcal{S}^{p_1-1,p_2-1,p_3-1}_{\Xi'_1,\Xi'_2,\Xi'_3},$$

where, for example, $\mathcal{S}^{p_1,p_2,p_3}_{\Xi_1,\Xi_2,\Xi_3}$ represents the space $\mathcal{S}^{p_1}_{\Xi_1} \otimes \mathcal{S}^{p_2}_{\Xi_2} \otimes \mathcal{S}^{p_3}_{\Xi_3}$, $\mathcal{S}^{p_1-1,p_2,p_3}_{\Xi'_1,\Xi_2,\Xi_3}$ represents the space $\mathcal{S}^{p_1-1}_{\Xi'_1} \otimes \mathcal{S}^{p_2}_{\Xi_2} \otimes \mathcal{S}^{p_3}_{\Xi_3}$, and so on. These spaces are suitable spaces for discrete 0-, 1-, 2- and 3-forms respectively, and they form a discrete de Rham complex,

$$\mathbb{R} \hookrightarrow \mathcal{S}^0_{\Xi_1,\Xi_2,\Xi_3} \xrightarrow{d} \mathcal{S}^1_{\Xi_1,\Xi_2,\Xi_3} \xrightarrow{d} \mathcal{S}^2_{\Xi_1,\Xi_2,\Xi_3} \xrightarrow{d} \mathcal{S}^3_{\Xi_1,\Xi_2,\Xi_3} \longrightarrow 0,$$

where the exterior derivative d has an exact, topological, sparse discrete counterpart, the incidence matrix. We call these spaces primal B-spline spaces and call the basis functions primal B-spline basis functions or simply primal B-splines.

2.2 Dual Representations

From now on, we restrict ourselves to $\mathbb{R}^3$.

Let $B^m_{i,j,k}$ be any primal B-spline in $\mathcal{S}^m_{\Xi_1,\Xi_2,\Xi_3}$ for $m = 0, 1, 2, 3$, then a basis dual functional $f^m_{r,s,t}$ satisfies the basic property

$$f^m_{r,s,t}\left(B^m_{i,j,k}\right) = \delta_{i,r}\delta_{j,s}\delta_{k,t}, \tag{3}$$

where δ is the Kronecker delta,

$$\delta_{i,r} = \begin{cases} 1 & \text{if } i = r \\ 0 & \text{otherwise} \end{cases}.$$

Dual functionals were already considered by de Boor [8]. A possible construction of a dual B-spline basis is through the use of the Gram matrix, see, for instance, Dornisch et al. [9]. Two elements $p_h^{(3)}$ and $q_h^{(3)}$ in $\mathcal{S}^3_{\Xi_1,\Xi_2,\Xi_3}$, for example, are represented by

$$p_h^{(3)}(\xi^1, \xi^2, \xi^3) = \sum_{i=1}^N \sum_{j=1}^N \sum_{k=1}^N p_{i,j,k} M^{i,p_1-1}_{\Xi'_1}(\xi^1) M^{j,p_2-1}_{\Xi'_2}(\xi^2) M^{k,p_3-1}_{\Xi'_3}(\xi^3)$$

and

$$q_h^{(3)}(\xi^1, \xi^2, \xi^3) = \sum_{i=1}^N \sum_{j=1}^N \sum_{k=1}^N q_{i,j,k} M^{i,p_1-1}_{\Xi'_1}(\xi^1) M^{j,p_2-1}_{\Xi'_2}(\xi^2) M^{k,p_3-1}_{\Xi'_3}(\xi^3). \tag{4}$$

The L^2-inner product of them is given by

$$\left(p_h^{(3)}, q_h^{(3)}\right)_{L^2(\Omega)} = \boldsymbol{p}^\mathsf{T} \mathrm{M}^{(3)} \boldsymbol{q}, \tag{5}$$

where $\boldsymbol{p}$ and $\boldsymbol{q}$ represent vectors of the expansion coefficients of $p_h^{(3)}$ and $q_h^{(3)}$ respectively. The matrix $\mathbb{M}^{(3)}$ is the mass matrix or the Gram matrix.

Instead of expanding $q^{(3)}$ using the primal B-splines in the way given in (4), we can expand it using basis functions which are linear combinations of the primal B-splines as long as these new basis functions are still linearly independent. For example, we can use a set of new basis functions, $\widetilde{\Phi}$, given by

$$\widetilde{\Phi} = \left(\mathbb{M}^{(3)}\right)^{-1}\Phi,$$

where $\Phi = \left(\cdots, M_{\Xi_1'}^{i,p_1-1}(\xi^1) M_{\Xi_2'}^{j,p_2-1}(\xi^2) M_{\Xi_3'}^{k,p_3-1}(\xi^3), \cdots\right)^{\mathrm{T}}$ is the vector of primal B-splines in $\mathcal{S}^3_{\Xi_1,\Xi_2,\Xi_3}$. Following directly the surjectivity and injectivity of $\mathbb{M}^{(3)}$, the proof of the linear independence of basis functions $\widetilde{\Phi}_i$ is trivial. We call basis functions $\widetilde{\Phi}_i$ the dual representations of primal B-splines Φ_j or simply the dual B-splines. The dual B-splines are bi-orthogonal to the primal B-splines in the sense that

$$\left(\widetilde{\Phi}_i, \Phi_j\right)_{L^2(\Omega)} = \delta_{i,j},$$

which mimics the duality pairing between dual basis functions, (3). Using these dual B-splines, a new expansion of $q^{(3)}$, expressed as $\widetilde{q}_h^{(3)}$, can be obtained. Its expansion coefficients are then given by

$$\widetilde{\boldsymbol{q}} = \mathbb{M}^{(3)}\boldsymbol{q}. \tag{6}$$

It is straightforward to see that, in this case,

$$\left(p_h^{(3)}, \widetilde{q}_h^{(3)}\right)_{L^2(\Omega)} = \boldsymbol{p}^{\mathrm{T}}\widetilde{\boldsymbol{q}}, \tag{7}$$

where, comparing to (5), effectively the mass matrix is removed from the inner product. And if we relate this to (2), we see that $\widetilde{q}_h^{(3)}$ plays the role of $\widetilde{\alpha}^k$ and $p_h^{(3)}$ represents w_α, and the mass matrix is an analogue of the metric tensor.

The main disadvantage of the conversion to dual B-splines is that one needs to multiply the primal B-splines with the inverse of the mass matrix and that the resulting dual B-splines no longer have local support. However, during the discretization, we do not explicitly set up these dual B-splines, but only use the property (7) in the weak formulation, which will give rise to very sparse systems. Afterwards, if we want to reconstruct the solutions, the dual B-splines have to be set up. This can be done locally for every patch in parallel, which is computationally efficient.

The same process of constructing dual B-splines can be applied to other primal B-spline spaces. Let $\widetilde{\mathcal{S}}^0_{\Xi_1,\Xi_2,\Xi_3}, \widetilde{\mathcal{S}}^1_{\Xi_1,\Xi_2,\Xi_3}, \widetilde{\mathcal{S}}^2_{\Xi_1,\Xi_2,\Xi_3}, \widetilde{\mathcal{S}}^3_{\Xi_1,\Xi_2,\Xi_3}$ denote the

$$\mathbb{R} \longrightarrow \mathcal{S}^0_{\Xi_1,\Xi_2,\Xi_3} \xrightarrow{d} \mathcal{S}^1_{\Xi_1,\Xi_2,\Xi_3} \xrightarrow{d} \mathcal{S}^2_{\Xi_1,\Xi_2,\Xi_3} \xrightarrow{d} \mathcal{S}^3_{\Xi_1,\Xi_2,\Xi_3} \longrightarrow 0$$

$$\Big\downarrow \mathbb{M}^{(0)} \qquad \Big\downarrow \mathbb{M}^{(1)} \qquad \Big\downarrow \mathbb{M}^{(2)} \qquad \Big\downarrow \mathbb{M}^{(3)}$$

$$0 \longleftarrow \widetilde{\mathcal{S}}^0_{\Xi_1,\Xi_2,\Xi_3} \xleftarrow{\widetilde{d}} \widetilde{\mathcal{S}}^1_{\Xi_1,\Xi_2,\Xi_3} \xleftarrow{\widetilde{d}} \widetilde{\mathcal{S}}^2_{\Xi_1,\Xi_2,\Xi_3} \xleftarrow{\widetilde{d}} \widetilde{\mathcal{S}}^3_{\Xi_1,\Xi_2,\Xi_3} \longleftarrow \mathbb{R}$$

Fig. 2 A discrete double de Rham complex of primal and dual B-spline spaces

dual B-spline spaces constructed from $\mathcal{S}^0_{\Xi_1,\Xi_2,\Xi_3}$, $\mathcal{S}^1_{\Xi_1,\Xi_2,\Xi_3}$, $\mathcal{S}^2_{\Xi_1,\Xi_2,\Xi_3}$, $\mathcal{S}^3_{\Xi_1,\Xi_2,\Xi_3}$ respectively. We can then construct the following discrete de Rham complex:

$$0 \longleftarrow \widetilde{\mathcal{S}}^0_{\Xi_1,\Xi_2,\Xi_3} \xleftarrow{\widetilde{d}} \widetilde{\mathcal{S}}^1_{\Xi_1,\Xi_2,\Xi_3} \xleftarrow{\widetilde{d}} \widetilde{\mathcal{S}}^2_{\Xi_1,\Xi_2,\Xi_3} \xleftarrow{\widetilde{d}} \widetilde{\mathcal{S}}^3_{\Xi_1,\Xi_2,\Xi_3} \hookleftarrow \mathbb{R},$$

where $\widetilde{d}$ is an adjoint exterior derivative acting on the dual spaces, see (1). For more information about $\widetilde{d}$, we refer to [12, 18]. Notice that a dual B-spline space is just a representation of its primal B-spline space. In other words, it is isomorphic to its primal B-spline space because the dual B-splines are just linear combinations of the primal B-splines. But here we still distinguish them because the L^2-inner product between an element in $\widetilde{\mathcal{S}}^k_{\Xi_1,\Xi_2,\Xi_3}$ and an element in $\mathcal{S}^k_{\Xi_1,\Xi_2,\Xi_3}$ mimics the duality pairing, see (7).

Finally, we can set up a particular instance, see Fig. 2, of the double de Rham complex in Fig. 1.

By now, all constructions in this paper are done in Cartesian domain. The construction of primal B-splines in curvilinear domains follows the general way of coordinates transformation, for example, see [17]. And the way of setting up their dual representations remains the same.

3 Numerical Test

In this section, we apply the B-spline spaces constructed in the previous section to a weak mixed formulation of the Poisson problem. Comparison between conditions of systems using the dual space and not using that is made.

3.1 Poisson Problem

In a bounded domain Ω in $\mathbb{R}^3$, a mixed formulation of the Poisson problem in differential forms is [16],

$$\begin{cases} u^{(2)} + d^*\varphi^{(3)} = 0 & \text{in } \Omega \\ du^{(2)} = -f^{(3)} & \text{in } \Omega \\ \text{tr} \star \varphi^{(3)} = \hat{\varphi}^{(0)} & \text{on } \partial\Omega \end{cases} \quad . \tag{8}$$

where $\hat{\varphi}^{(0)}$ and $f^{(3)}$ are given, d^*, the codifferential operator, is the adjoint of the exterior derivative d in terms of the L^2-inner product, and $\star$ is the Hodge operator. A weak formulation of (8) is then given as: For known $\left(\hat{\varphi}^{(0)}, f^{(3)}\right) \in H^{1/2}\Lambda^0(\partial\Omega) \times L^2\Lambda^3(\Omega)$, find $\left(u^{(2)}, \varphi^{(3)}\right) \in H^1\Lambda^2(\Omega) \times L^2\Lambda^3(\Omega)$, such that

$$\begin{cases} \left(u^{(2)}, \bar{u}^{(2)}\right)_{L^2(\Omega)} + \left(\varphi^{(3)}, d\bar{u}^{(2)}\right)_{L^2(\Omega)} = \int_{\partial\Omega} \text{tr}\,\bar{u}^{(2)} \wedge \hat{\varphi}^{(0)} & \forall \bar{u}^{(2)} \in H^1\Lambda^2(\Omega) \\ \left(du^{(2)}, \bar{\varphi}^{(3)}\right)_{L^2(\Omega)} = -\left(f^{(3)}, \bar{\varphi}^{(3)}\right)_{L^2(\Omega)} & \forall \bar{\varphi}^{(3)} \in L^2\Lambda^3(\Omega) \end{cases}$$

To discretize this weak formulation, a conventional choice is to select finite dimensional spaces $\mathcal{S}^2_{\Xi_1,\Xi_2,\Xi_3} \times \mathcal{S}^3_{\Xi_1,\Xi_2,\Xi_3}$ for $H^1\Lambda^2(\Omega) \times L^2\Lambda^3(\Omega)$. This gives rise to the following system,

$$\begin{bmatrix} \mathbb{M}^{(2)} & \left(\mathbb{E}^{3,2}\right)^T \mathbb{M}^{(3)} \\ \mathbb{M}^{(3)}\mathbb{E}^{3,2} & 0 \end{bmatrix} \begin{bmatrix} u \\ \varphi \end{bmatrix} = \begin{bmatrix} \mathbb{B}\hat{\varphi} \\ -\mathbb{M}^{(3)}f \end{bmatrix}, \tag{9}$$

where $\mathbb{M}^{(2)}$ and $\mathbb{M}^{(3)}$ are mass matrices of spaces $\mathcal{S}^2_{\Xi_1,\Xi_2,\Xi_3}$ and $\mathcal{S}^3_{\Xi_1,\Xi_2,\Xi_3}$, $\mathbb{E}^{3,2}$ is the incidence matrix which represents the discrete exterior derivative d : $\mathcal{S}^2_{\Xi_1,\Xi_2,\Xi_3} \to \mathcal{S}^3_{\Xi_1,\Xi_2,\Xi_3}$ and $\mathbb{B}$ represents the matrix generated from the boundary integral term. We call this setup the primal-primal setup and we denote the left hand side matrix by $\mathbb{P}$.

We can alternatively use the primal-dual setup in which we select spaces $\mathcal{S}^2_{\Xi_1,\Xi_2,\Xi_3} \times \widetilde{\mathcal{S}}^3_{\Xi_1,\Xi_2,\Xi_3}$ for $H^1\Lambda^2(\Omega) \times L^2\Lambda^3(\Omega)$. By doing this, the discrete system will be

$$\begin{bmatrix} \mathbb{M}^{(2)} & \left(\mathbb{E}^{3,2}\right)^T \\ \mathbb{E}^{3,2} & 0 \end{bmatrix} \begin{bmatrix} u \\ \tilde{\varphi} \end{bmatrix} = \begin{bmatrix} \mathbb{B}\hat{\varphi} \\ -f \end{bmatrix}. \tag{10}$$

We denote the left hand side matrix of this system by $\mathbb{D}$.

It is easy to see that systems (9) and (10) are equivalent in terms of linear algebra if we replace $\widetilde{\varphi}$ by $\mathbb{M}^{(3)}\varphi$, see (6). However, the system of the primal-dual setup is easier to construct, and matrices $\mathbb{P}$ and $\mathbb{D}$ have different conditions, for instance sparsity and condition number, which will affect their solvabilities.

3.2 Manufactured Solution Test Case

To study the convergence of the method and the conditions of $\mathbb{P}$ and $\mathbb{D}$, we do a test with a manufactured solution in both Cartesian and curvilinear domains in $\mathbb{R}^3$.

The Cartesian domain is selected to be a unit cube, $x = y = [-0.5, 0.5]$, $z = [0, 1]$. The curvilinear domains are obtained by fixing the bottom surface, $z = 0$, of the unit cube and twisting it along the z-axis. In polar coordinates, the mapping is described by

$$\begin{cases} r_t = r \\ \theta_t = \theta + tz \\ z_t = z \end{cases},$$

where the deformation coefficient, t, represents the maximum twist on the upper surface, $z = 1$. We use Ω_t to express such a domain. Some examples of Ω_t are shown in Fig. 3. The whole domain is considered as one patch and N^3 elements are uniformly distributed. Within these domains, we solve the system (10) with the following manufactured solution,

$$\varphi_{\text{exact}}^{(0)} = \star \varphi_{\text{exact}}^{(3)} = \cos(\pi x y) e^z .$$

The B-spline degree is selected to be p along all three axes.

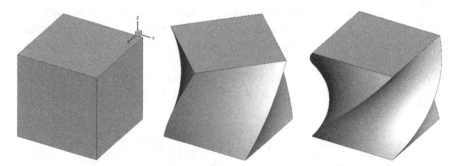

Fig. 3 The Cartesian domain (unit cube) Ω_0 (Left) and curvilinear domains $\Omega_{\pi/4}$ (Middle) and $\Omega_{\pi/2}$ (Right)

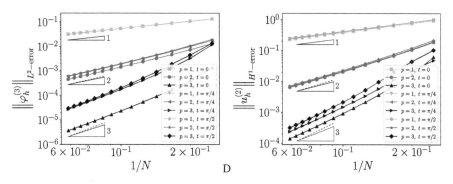

Fig. 4 The L^2-error of $\varphi_h^{(3)}$ and the H^1-error of $u_h^{(2)}$ under a h-refinement

The implementation is done within Python and the system is solved with the direct solver provided by the Scipy package. Iterative refinement [15] is employed to reduce the roundoff error.

We first investigate the convergence of the method. In Fig. 4, it is seen that the optimal algebraic convergence is always obtained for the L^2-error of $\varphi_h^{(3)}$ and the H^1-error of $u_h^{(2)}$ when we do h-refinement in both Cartesian and curvilinear domains. The H^1-error, a generalization of the $H(\text{grad})$-, $H(\text{curl})$- and $H(\text{div})$-error in vector calculus, is defined as

$$\left\| a_h^{(k)} \right\|_{H^1-\text{error}} = \sqrt{\left\| a_h^{(k)} \right\|^2_{L^2-\text{error}} + \left\| \mathrm{d} a_h^{(k)} \right\|^2_{L^2-\text{error}}}.$$

In Fig. 5 where results of the L^2-error of $\varphi_h^{(3)}$ and the H^1-error of $u_h^{(2)}$ under a p-refinement are presented, we see that the exponential convergence is only obtained for a limited range of p. This is because of the rapidly increasing condition number of the system $\mathbb{D}$ damages the accuracy of the solver at high B-spline degree, see,

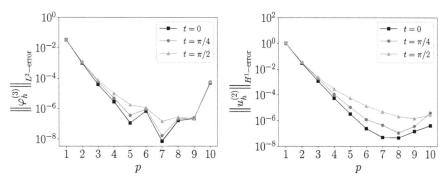

Fig. 5 The L^2-error of $\varphi_h^{(3)}$ and the H^1-error of $u_h^{(2)}$ under a p-refinement

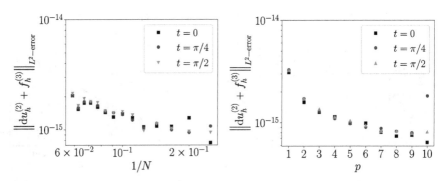

Fig. 6 The L^2-error of $\left(\mathrm{d}u_h^{(2)} + f_h^{(3)}\right)$ under a h-refinement (Left) and under a p-refinement (Right)

for example, the L^2-error of $\varphi_h^{(3)}$ at $p = 10$. An investigation into the condition numbers will be given later. In Fig. 6, we present the L^2-error of $\left(\mathrm{d}u_h^{(2)} + f_h^{(3)}\right)$. It shows that the relation $\mathrm{d}u_h^{(2)} = -f_h^{(3)}$, the mass conservation, is always preserved to the machine precision, which proves the discretization of the exterior derivative d with the incidence matrix is exact.

Since the systems (9) and (10) are equivalent in terms of linear algebra, besides the fact that (10) is easier to construct, we expect the gain of using dual B-splines is that $\mathbb{D}$ has better condition than $\mathbb{P}$. Most obviously, the sparsity of $\mathbb{D}$ can be higher than that of $\mathbb{P}$ since the mass matrix $\mathbb{M}^{(3)}$ is removed from the system. We have presented the sparsities of $\mathbb{P}$ and $\mathbb{D}$ for $N = 3$ and $t = 0$ (Cartesian domain) in Fig. 7. In this figure, we can see that, when $p = 1$, $\mathbb{P}$ and $\mathbb{D}$ have the same sparsity because $\mathbb{M}^{(3)}$ is a diagonal matrix in this case. When we increase p, the sparsity of $\mathbb{M}^{(3)}$ decreases, which then results in a higher sparsity in $\mathbb{D}$ than that in $\mathbb{P}$. And when the domain is curvilinear, the sparsity of $\mathbb{M}^{(3)}$ is even worse (if $p < N$, otherwise, $\mathbb{M}^{(3)}$ is a full matrix anyway). We emphasize that the incidence matrix only depends on the topology of the mesh. So it does not change when we change the B-spline degree p or distort the domain. We conclude that, in terms of sparsity, the primal-dual setup is preferable.

We then investigate the condition numbers of $\mathbb{P}$ and $\mathbb{D}$. The results are given in Fig. 8. It has shown that for given B-spline degrees p, see Fig. 8a–c, the matrix $\mathbb{D}$ has much lower condition numbers than $\mathbb{P}$ on coarse meshes. And when we refine the mesh, the condition numbers of $\mathbb{P}$ grow steadily while the condition numbers of $\mathbb{D}$ first almost remain unchanged (apparently for $p \geq 2$). However, after a certain mesh density, N_0, condition numbers of $\mathbb{D}$ increase even faster than those of $\mathbb{P}$. The

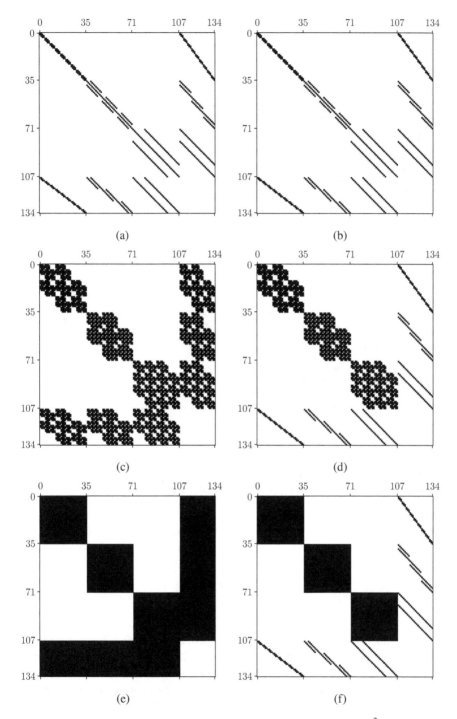

Fig. 7 Sparsities of $\mathbb{P}$ (Left) and $\mathbb{D}$ (Right) for $N = 3$ and $t = 0$. The $[0:107]^2$ block refers to the upper-left block, $\mathbb{M}^{(2)}$, of $\mathbb{P}$ and $\mathbb{D}$. (**a**, **b**) $p = 1$. (**c**, **d**) $p = 2$. (**e**, **f**) $p = 3$

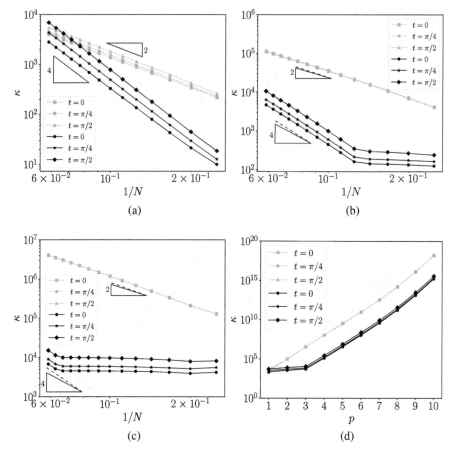

Fig. 8 Condition numbers of the primal-primal system $\mathbb{P}$ (gray) and the primal-dual system $\mathbb{D}$ (black). (**a**) $p = 1$. (**b**) $p = 2$. (**c**) $p = 3$. (**d**) $N = 16$

transition point N_0 depends on p. When p increases, N_0 will be delayed:

$$N_0 \leq 4 \quad \text{when } p = 1,$$
$$N_0 = 8 \quad \text{when } p = 2,$$
$$N_0 = 15 \quad \text{when } p = 3.$$

When we do p-refinement, a similar phenomenon is observed. But the transition point does not depend on the mesh density. After the transition, the increasing ratio of $\mathbb{D}$'s condition numbers is the same with that of $\mathbb{P}$'s, for example, see Fig. 8d. To fully understand this phenomenon, more research is needed. In the current paper, we will just cautiously say that, considering the most widely used B-splines are those of degree $p = 3$ or 4, the dual B-splines, under careful usage, can bring down the condition numbers of the discrete systems by several magnitudes.

4 Conclusions

In this paper, we propose a way of constructing dual representations of B-spline basis functions. The spaces spanned by these B-splines and their dual representations form a discrete double de Rham complex, with which exact discretization of the exterior derivative can be obtained. The dual B-spline spaces, like preconditioners, can give rise to much more sparse discrete systems and can conditionally decrease the condition numbers significantly.

References

1. Arnold D., Falk R., Winther R.: Finite element exterior calculus: from Hodge theory to numerical stability. Bulletin of the American Mathematical Society, **47(2)**, 281–354, (2010)
2. Back A., Sonnendrücker E.: Spline discrete differential forms. Application to Maxwell's equations. hal-00568811, (2011)
3. Bochev, P.B., Hyman J.M.: Principles of mimetic discretizations of differential operators. Compatible spatial discretizations. Springer, New York, NY, 89–119, (2006)
4. Buffa A., Rivas J., Sangalli G., Vázquez R.: Isogeometric discrete differential forms in three dimensions. *SIAM J. Numer. Anal.*, **49(2)**, 818–844, (2011)
5. Buffa A., Sangalli G., Vázquez R.: Isogeometric methods for computational electromagnetics: B-spline and T-spline discretizations. *Journal of Computational Physics*, **257**, 1291–1320, (2014)
6. Cottrell J.A., Hughes T.J.R., Bazilevs Y.: Isogeometric analysis: towards integration of CAD and FEA. John Wiley & Sons, Ltd, (2009)
7. de Boor C., Fix G.J.: Spline approximation by quasiinterpolants. *Journal of Approximation Theory*, **8(1)**, 19–45, (1973)
8. de Boor C.: On local linear functionals which vanish at all B-splines but one. Mathematics Research Center, University of Wisconsin, (1975)
9. Dornisch W., Stöckler J., Müller R.: Dual and approximate dual basis functions for B-splines and NURBS – Comparison and application for efficient coupling of patches with the isogeometric mortar method. Comput. *Methods Appl. Mech. Engrg.*, **316**, 449–496, (2017)
10. Evans J.A., Scott M.A., Shepherd K., Thomas D., Vázquez R.: Hierarchical B-spline complexes of discrete differential forms. arXiv:1708.04195, (2017)
11. Hiemstra R., Toshniwal D., Huijsmans R.H.M., Gerritsma M.: High order geometric methods with exact conservation properties. *Journal of Computational Physics*, **257**, 1444–1471, (2014)
12. Jain V., Zhang Y., Palha A., Gerritsma M.: Construction and application of algebraic dual polynomial representations for finite element methods. arXiv:1712.09472, (2017)
13. Kreeft J., Palha A., Gerritsma M.: Mimetic framework on curvilinear quadrilaterals of arbitrary order. arXiv preprint arXiv:1111.4304, (2011)
14. Kreyszig E.: Introductory Functional Analysis with Applications. Vol. 1. New York: Wiley, (1978)
15. Moler C.B.: Iterative refinement in floating point. *Journal of the ACM (JACM)*, **14(2)**, 316–321, (1967)

16. Palha A., Rebelo P.P., Hiemstra R., Kreeft J., Gerritsma M.: Physics-compatible discretization techniques on single and dual grids, with application to the Poisson equation of volume forms. *Journal of Computational Physics*, **257**, 1394–1422, (2014)
17. Steinberg S.: Fundamentals of grid generation. CRC Press, (1993)
18. Zhang Y., Jain V., Palha A., Gerritsma M.: Discrete equivalence of adjoint Neumann-Dirichlet div-grad and grad-div equations in curvilinear 3D domains. ICOSAHOM2018 conference proceeding, (2018)

Manifold-Based B-Splines on Unstructured Meshes

Qiaoling Zhang, Thomas Takacs, and Fehmi Cirak

Abstract We introduce new manifold-based splines that are able to exactly reproduce B-splines on unstructured surface meshes. Such splines can be used in isogeometric analysis (IGA) to represent smooth surfaces of arbitrary topology. Since prevalent computer-aided design (CAD) models are composed of tensor-product B-spline patches, any IGA suitable construction should be able to reproduce B-splines. To achieve this goal, we focus on univariate manifold-based constructions that can reproduce B-splines. The manifold-based splines are constructed by smoothly blending together polynomial interpolants defined on overlapping charts. The proposed constructions are able to reproduce B-splines in regular parts of the mesh, with no extraordinary vertices, and polynomial basis functions in the remaining parts of the mesh. We study and compare analytically and numerically the finite element convergence of several univariate constructions. The obtained results directly carry over to the tensor-product case.

1 Introduction

Manifold-based surface construction techniques from geometric modelling provide an elegant and flexible framework for generating basis functions on surfaces with arbitrary topology [9, 16, 25, 8]. They combine manifold descriptions from differential geometry, see e.g. [20, 3], with the flexibility of the partition of unity framework from numerical analysis [15, 1]. If a manifold surface in $\mathbb{R}^3$ can be mapped onto a single planar parametric domain in $\mathbb{R}^2$, it is straightforward to obtain partition of unity basis functions of any desired regularity on the manifold surface.

Q. Zhang · F. Cirak (✉)
Department of Engineering, University of Cambridge, Cambridge, UK
e-mail: zq217@cam.ac.uk; fc286@cam.ac.uk

T. Takacs
Institute of Applied Geometry, Johannes Kepler University, Linz, Austria
e-mail: thomas.takacs@jku.at

Although it is impossible to map a surface with arbitrary topology onto a single parametric domain, it can always be represented as an atlas composed of a number of charts. The charts consist of planar domains in $\mathbb{R}^2$ that map onto the manifold surface in $\mathbb{R}^3$. The planar chart domains in $\mathbb{R}^2$ are not connected and transition functions are used to navigate between the different domains. The manifold-based basis functions are obtained by simply applying the partition of unity method on the collection of chart domains [14, 26]. The flexibility of the original partition of unity method carries over to the manifold case. The partition of unity functions, referred to as blending functions, in this paper, the local approximants on each chart domain and the transition functions can all be chosen to fit the requirements of the application at hand.

The definition of smooth functions over unstructured meshes, such as shown in Fig. 1, or multi-patch geometries has always been of vital interest in isogeometric analysis, cf. [10, 7]. Approaches for the definition of such functions include the subdivision surfaces [5, 4, 18, 27], constructions that are C^k almost everywhere [19, 2], G^k constructions [21, 22, 17, 6, 11, 12, 13] and C^k constructions with singular parameterisations [24, 23]. Since most conventional CAD models are based on B-spline or NURBS surfaces, any isogeometric analysis suitable construction should be able to reproduce tensor-product B-splines and NURBS. For this reason we especially focus on manifold-based constructions that can reproduce tensor-product B-splines in regular portions of the mesh.

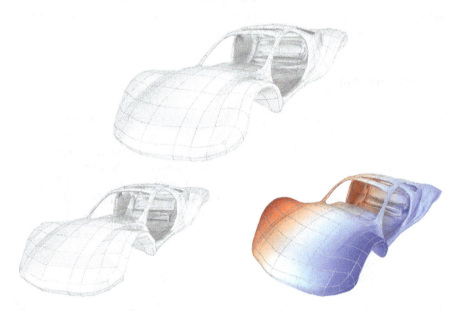

Fig. 1 Isogeometric analysis of a car body as used in computer animation. The manifold-based representation (top) is obtained from an unstructured coarse quadrilateral control mesh (bottom, left), with extraordinary vertices indicated by blue spheres. The deflected shape of the car body subjected to an axial torsion is computed with Kirchhoff-Love shell finite elements (bottom, right)

In this paper, we exploit the flexibility of the partition of unity method to devise manifold-based basis functions that can reproduce or are identical to B-splines. The proposed techniques are introduced for the sake of clarity with the help of univariate B-splines. Evidently, manifold-based basis functions can reproduce B-splines only on structured regions of an unstructured mesh with extraordinary vertices, i.e. non-boundary vertices with different than four attached elements. In the vicinity of extraordinary vertices, the basis functions consist of a local polynomial approximant that can smoothly blend with the surrounding B-spline reproducing basis functions. The extent of the transition region depends on the size of the chosen chart domain, which consists of an n_v-ring of elements around each vertex. We consider several different choices for the weight functions that lead to B-spline reproducing basis functions. Especially promising are weight functions which are defined as a linear combination of B-splines defined on a grid obtained by subdividing the elements multiple times. They satisfy partition of unity without normalisation and, hence, lead to polynomial manifold-based basis functions. To obtain manifold-basis functions that are identical to B-splines the local polynomial approximants have to be altered. Whereas in the original manifold constructions the local polynomial approximants are C^∞, they have to be chosen to have the same smoothness as the considered B-splines.

On structured meshes the approximation properties of manifold-based basis functions can be inferred from the theory presented in Melenk and Babuska [15]. The summation of the local errors on the charts gives a global error estimate under some smoothness assumptions on the weight functions. The local error, for instance in L^∞ or L^2 norms, is bounded by h^{p+1}, where h is the diameter of the chart domain and p is the degree of the polynomials contained in the local approximant. Each chart domain consists of n_v-rings of elements around a vertex so that there are on unstructured meshes multiple types of chart domains depending on the local connectivity of the control mesh. The local connectivity of the control mesh determines the type of transition function used in the manifold construction. When a control mesh is refined by quadrisecting its elements, all the newly introduced vertices are ordinary. That is, for points close to the extraordinary vertices the type of the transition function used depends on the refinement level of the mesh. Therefore, the theory presented in Melenk and Babuska [15] has to be extended to cover the extraordinary vertices, which we do not attempt in this paper.

The outline of this paper is as follows. In Sect. 2 we review the manifold basis functions as introduced in [14]. Although only univariate basis functions on polygonal control meshes are considered, no specific choices for the weight functions, local approximants and the transition functions are given so that the presented theory is applicable to the multivariate case as well. Subsequently, in Sect. 3 several specific choices, first for weight functions and then for local approximants are introduced. More specifically, in Sect. 3.1 five choices for the weight functions are proposed, two of which yield manifold basis functions that can reproduce B-splines. In Sect. 3.2 it is illustrated how to choose the local approximants so that manifold-basis functions are identical to B-splines. Finally, in Sect. 4 we provide a summary and comparison of the different proposed constructions.

2 Review of Manifold-Based Basis Functions

We provide a brief informal review of univariate manifold basis functions for curves with the aim to fix ideas and notation.

2.1 Basic Approach

Given is a control polygon with the vertex coordinates $\mathbf{x}_i \in \mathbb{R}^3$ which describes the manifold curve Ω. To begin with, the control polygon and the curve are assumed to be closed to sidestep the discussion of boundaries. The curve is composed of a set of n_c overlapping subdomains

$$\Omega = \bigcup_{i=1}^{n_c} \Omega_i. \tag{1}$$

Each subdomain Ω_i is associated with a vertex $\mathbf{x}_i$ of the control polygon in a manner yet to be described. The subdomains Ω_i are obtained from corresponding planar domains $\hat{\Omega}_i \in \mathbb{R}$ with a mapping

$$\begin{aligned} \varphi_i : \hat{\Omega}_i &\to \Omega_i \\ \xi_i &\mapsto \mathbf{x}. \end{aligned} \tag{2}$$

The pair consisting of $(\hat{\Omega}_i, \varphi_i)$ is called a chart. In the following we refer to $\hat{\Omega}_i$ as the chart domain or simply as the chart. Each chart domain $\hat{\Omega}_i$ has its own coordinate system with the coordinates $\xi_i \in \mathbb{R}$. The coordinates of points on the intersection between two subdomains Ω_i and Ω_j can be mapped with transition functions, that is,

$$\begin{aligned} t_{ji} : \hat{\Omega}_i^j \subset \hat{\Omega}_i &\to \hat{\Omega}_j \\ \xi_i &\mapsto \xi_j \end{aligned} \tag{3}$$

defined as

$$t_{ji} = \varphi_j^{-1} \circ \varphi_i. \tag{4}$$

Here, $\hat{\Omega}_i^j = \varphi_i^{-1}(\Omega_i \cap \Omega_j)$ is the pull-back of the intersection of the two subdomains.

For constructing a smooth approximant, on each chart domain we have given a blending (or, weight) function $w_i : \hat{\Omega}_i \to \mathbb{R}_0^+$ with

$$\mathrm{supp}(w_i) \subseteq \hat{\Omega}_i$$

which have to satisfy

$$\sum_{i=1}^{n_c} w_i \circ \varphi_i^{-1} \equiv 1 \quad \text{on } \Omega \tag{5}$$

and have to be C^k smooth. In addition, at the chart domain boundaries, w_i and its derivatives up to k-th order have to be zero. On each chart domain also a local approximant $f_i : \hat{\Omega}_i \to \mathbb{R}$ is defined. The approximant f_i is usually expressed in a polynomial basis, like the power, Lagrangian or the Bézier basis. In this paper both the Lagrangian and Bézier basis of a fixed degree are used. However, the basis and the degree of the approximant f_i may be different on every chart. Hence, having the local bases $\mathcal{P}_i = \left\{ p_i^{(j)}(\xi_i) \right\}_{j=1}^{q_p+1}$ of degree q_p and the corresponding local coefficients $\boldsymbol{\alpha}_i = \{\alpha_i^{(j)}\}$ on chart $\hat{\Omega}_i$ gives the local approximant

$$f_i(\xi_i) = \sum_{j=1}^{q_p+1} p_i^{(j)}(\xi_i) \alpha_i^{(j)} := \mathbf{p}_i^T(\xi_i) \boldsymbol{\alpha}_i, \tag{6}$$

and, in turn, the global approximant

$$f(\xi_i) = \sum_{l:\, \varphi_l(\xi_l) = \varphi_i(\xi_i)} w_l(\xi_l) f_l(\xi_l) = \sum_{l:\, \varphi_l(\xi_l) = \varphi_i(\xi_i)} w_l(\xi_l) \mathbf{p}_l^T(\xi_l) \boldsymbol{\alpha}_l, \tag{7}$$

as well as the global basis

$$\mathcal{P}_{global} = \bigcup_{i=1}^{n_c} w_i \mathcal{P}_i = \{ w_i(\xi_i) p_i^{(j)}(\xi_i) : \text{ with } i = 1, \ldots, n_c \text{ and } j = 1, \ldots, q_p + 1 \}. \tag{8}$$

Note that the index i from the basis $p_i^{(j)}$ may be dropped when on each chart domain $\hat{\Omega}_i$ the same basis is used, as in the present paper.

Next, each chart domain $\hat{\Omega}_i$ and its image Ω_i are associated with segments/elements in the n_v-neighbourhood of the vertex $\mathbf{x}_i$ of the given control polygon. That is, there are as many charts as vertices in the mesh. An 1-neighbourhood of a vertex is defined as the union of elements that contain the vertex. The n_v-neighbourhood is defined recursively as the union of all 1-neighbourhoods of the $(n_v - 1)$-neighbourhood vertices. The number of segments associated with a chart is hence $2n_v$. In turn, each segment is present in $2n_v$ charts. See Fig. 2 for a construction where each chart domain $\hat{\Omega}_i$ consists of the two segments in the 1-neighbourhood of the vertex $\mathbf{x}_i$.

In Sect. 3 we consider the span of $\mathcal{P}_{global}$ in (8) as the *analysis space* on Ω. That is, the basis together with the corresponding set of coefficients $\{\boldsymbol{\alpha}_j\}$ is used

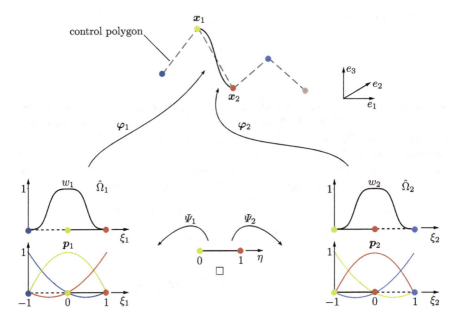

Fig. 2 Construction of a univariate manifold basis. The chart domains $\hat{\Omega}_1$ and $\hat{\Omega}_2$ are chosen to consist out of two segments. The segment $[\mathbf{x}_1, \mathbf{x}_2]$ on the (dashed) control polygon is present on both chart domains. On each chart the local basis $\mathbf{p}_1(\xi_1)$ and $\mathbf{p}_2(\xi_2)$ is a quadratic Lagrange basis and the blending functions $w_1(\xi_1)$ and $w_2(\xi_2)$ are normalised B-splines constructed according to (16), see also Fig. 6. The reference element (i.e. reference finite element) is denoted with $\square$. The same vertex has the same colour in the four domains $\square$, $\hat{\Omega}_1$, $\hat{\Omega}_2$ and Ω. Note that for the considered quadratic Lagrange basis and three vertices per chart domain the least-squares projection matrices $\mathbf{A}_1$ and $\mathbf{A}_2$ are identity matrices. Only choosing a constant or linear Lagrange basis on each chart domain requires a least-squares projection

for L^2-fitting. However, the set of coefficients $\{\boldsymbol{\alpha}_j\}$ lack an intuitive interpretation, similar to the control vertices of splines, so that the basis (8) is not suitable for geometric modelling. Hence, in the following we define a *design space* as a suitable subspace having degrees of freedom corresponding to the vertices $\mathbf{x}_i$.

2.2 Mesh-Based Approach

On each chart domain the coefficients of the local approximant can be assigned to vertices in the n_v-neighbourhood, see Fig. 2. Each vertex is present on $2n_v + 1$ charts which leads to a coupling between the coefficients of the local approximants (6) of the involved charts. If the number of the coefficients of the local approximant is less than the number of vertices in the chart a least squares fitting has to be applied

$$\boldsymbol{\alpha}_i = \mathbf{A}_i \mathbf{P}_i \mathbf{f}, \qquad (9)$$

where $\mathbf{A}_i$ denotes the least-squares projection matrix, $\mathbf{f}$ is an array of the scalar vertex coefficients $f_i \in \mathbb{R}$ for the entire polygon (one scalar per vertex) and $\mathbf{P}_i$ a gather matrix filled with ones and zeros to pick up the control vertex coefficients for a particular chart from $\mathbf{f}$. Hence, the global approximant (7) can be rewritten as

$$f(\xi_i) = \sum_{l:\, \varphi_l(\xi_l) = \varphi_i(\xi_i)} w_l(\xi_l) \mathbf{p}_l^T(\xi_l) \mathbf{A}_l \mathbf{P}_l \mathbf{f}. \tag{10}$$

Finally, the manifold curve Ω can be obtained by replacing the array of vertex scalar coefficients $\mathbf{f}$ with the array of given vertex coordinates $\mathbf{x}$, so that each map (2) reads

$$\varphi_i(\xi_i) = w_i(\xi_i) \mathbf{p}_i^T(\xi_i) \mathbf{A}_i \mathbf{P}_i \mathbf{x}. \tag{11}$$

To summarise so far, each segment on the control mesh has a unique set of corresponding segments on several planar chart domains $\hat{\Omega}_i$. The introduced manifold construction ensures that the images of the set of segments from disparate planar charts are identical on the manifold Ω. To advance a more classical finite element interpretation, each segment on the manifold (Ω, s) with the index s represents an element and has a corresponding reference element $(\Box, s)$ to evaluate the element integrals, where $\Box := [0, 1]$.[1] The mapping of the parent element onto the manifold is composed of two maps

$$\begin{array}{c} \varphi_i \circ \Psi_{i,s} : (\Box, s) \to (\hat{\Omega}_i, s) \to (\Omega, s) \\ (\eta, s) \mapsto \xi_i \mapsto \mathbf{x} \end{array} \tag{12}$$

with $(\hat{\Omega}_i, s) \subset \hat{\Omega}_i$ and $(\Omega, s) \subset \Omega$ being a segment on the chart domain or manifold, respectively. This implies for the field variables in a reference element s,

$$f(\eta, s) \equiv f(\Psi_{i,s}^{-1}(\varphi_i^{-1}(\mathbf{x}))). \tag{13}$$

In applications the maps $\Psi_{i,s}$ have to be chosen carefully. Namely, the transition functions defined in (4) can be determined as, c.f. Fig. 2,

$$t_{ji} = \Psi_{j,s} \circ \Psi_{i,s}^{-1} \tag{14}$$

so that the required smoothness of t_{ji} depends on the collection of maps $\Psi_{i,s}$ and $\Psi_{j,s}$ on the respective chart domains.

---

[1] The index s for the reference element $\Box$ is usually dropped because all of them can be assumed to have the same domain.

The approximation of the field variables with (10) leads for the element s to the following definition of finite element basis functions $\mathbf{N}(\eta, s)$:

$$f(\eta, s) = \underbrace{\left(\sum_{j:\, \varphi_j(\xi_j) = \varphi_i(\xi_i)} w_j(\xi_j) \mathbf{p}_j^\mathsf{T}(\xi_j) \mathbf{A}_j \mathbf{P}_j \right)}_{\mathbf{N}^\mathsf{T}(\eta, s)} \mathbf{f} \quad \text{with} \quad \xi_j = \Psi_{j,s}(\eta). \tag{15}$$

In the following we denote the basis functions with $\mathbf{N}(\eta)$ and the mapping from the reference element to the chart with Ψ_s. This notation is not precise when charts have different geometries and number of vertices.

It is clear that the smoothness of the basis functions $\mathbf{N}(\eta)$ depends on the smoothness of the blending functions $w_i(\xi_i)$, the local basis $\mathbf{p}_i(\xi_i)$ and the mappings $\Psi_i(\eta)$. For instance, in the two-dimensional C^k continuous construction introduced in [14], the blending functions w_i are chosen to be (normalised) B-splines of degree $k+1$, the local basis $\mathbf{p}_i$ are chosen to be a polynomial basis and Ψ_i are conformal maps. Furthermore, each chart domain consists of the elements in the 1-neigborhood of the corresponding vertex. Figure 2 illustrates this construction for C^2 continuous basis functions in the univariate case.

3 Reproduction of B-Splines

We consider again the global basis (8) for analysis, repeated here for convenience,

$$\mathcal{P}_{global} = \bigcup_{i=1}^{n_c} w_i \mathcal{P}_i = \{ w_i(\xi_i) p_i^{(j)}(\xi_i) :\ \text{with } i = 1, \ldots, n_c \text{ and } j = 1, \ldots, q_p + 1 \},$$

and discuss how to choose the blending functions w_i and the local basis $p_i^{(j)}$ so that B-splines are a subset of the basis $\mathcal{P}_{global}$. To reproduce B-splines it is sufficient to choose either the blending functions w_i or the local basis $\mathbf{p}_i$ suitably. In the following we introduce several choices for the blending functions and local basis and comment on their extendability to the bivariate case. Evidently, B-splines are defined on a structured mesh so that manifold-based basis functions will only reproduce B-splines on the parts of the mesh with no extraordinary vertices.

In the univariate case, the parameter domain of the manifold curve Ω can be assumed to be one single finite interval $\hat{\Omega}$. Due to the choice of the single finite interval the transition functions t_{ij} are identity maps and Ψ_s are affine maps, both are omitted in the following. Without loss of generality, the parameter domain is uniformly partitioned with n inner nodes with the coordinates $\hat{\xi}_j = j$. Moreover, we use the notation $\mathbb{P}^p$ for polynomials of degree p, $\mathcal{S}^{p,r}$ for B-splines of degree p and continuity C^r and define the space

$$\mathcal{W} = \text{span}\{w_i : i = 1, \ldots, n_c\}$$

as a function space on $\hat{\Omega}$.

3.1 Blending Function Choices

We fix the local basis $p_i^{(j)}$ to be polynomials of some prescribed degree and study how to choose the blending functions w_i to obtain B-spline reproducing manifold-based basis functions. As the B-splines form a local, non-negative partition of unity, they are used as blending functions. Specifically, the blending functions will be chosen either as

- standard B-splines of maximum smoothness,
- rational B-spline functions, or
- linear combinations of B-splines.

We compare the different approaches in terms of maximum number of overlapping charts at any point of the domain, number of degrees of freedom, expected approximation order as well as smoothness properties.

3.1.1 Piecewise Linear C^0 Continuous Blending Functions

In case of linear B-spline blending functions of degree $q_w = 1$ each chart domain contains three knots. This means that any point on the parameter domain is present on two different chart domains. Having C^0 hat functions as blending functions and polynomials of degree q_p as local functions on every chart, we reproduce continuous, piecewise polynomials of degree $q_p + 1$, i.e., C^0 B-splines. Hence, $\mathcal{W} = \mathcal{S}^{1,0}$ and $\text{span}(\mathcal{P}_{global}) = \mathcal{S}^{q_p+1,0}$. In Fig. 3 the manifold-based basis obtained with linear B-spline blending functions and a cubic Bézier local basis is shown. The corresponding hat function and the cubic Bézier basis are depicted in Fig. 4.

Extending the construction to surfaces, we obtain functions that are tensor-product polynomials within every quadrilateral element and C^0 continuous across every edge. Hence, we can only reproduce C^0 continuous basis functions on unstructured quadrilateral Bézier meshes.

Fig. 3 Global basis for piecewise linear blending function with $q_w = 1$ and local cubic Bézier basis with $q_p = 3$ restricted to the chart domain $\hat{\Omega}_i = \left[\hat{\xi}_{i-1}, \hat{\xi}_{i+1}\right]$

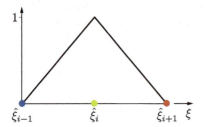 

Fig. 4 Piecewise linear blending function with $q_w = 1$ (left) and a local cubic Bézier basis with $q_p = 3$ (right) used in computing the basis in Fig. 3

3.1.2 Higher Order C^{p-1} Continuous B-Spline Blending Functions

The generalisation of the linear B-spline blending functions to the higher order B-splines with $q_w > 1$ is straightforward. As shown in Fig. 5 the support of each basis function on the parameter domain is defined as a chart. Therefore, for B-splines of degree q_w and smoothness $q_w - 1$ in any point of the domain $q_w + 1$ charts overlap.

Taking splines of degree $q_w > 1$ as blending functions, together with polynomials of degree q_p as local basis, we reproduce B-splines of degree $q_w + q_p$ and smoothness $q_w - 1$. Hence, we have $\mathcal{W} = \mathcal{S}^{q_w, q_w-1}$. Let $\mathcal{W}_0$ be the subspace of $\mathcal{W}$ without global polynomials, i.e.,

$$\mathcal{W}_0 = \mathcal{W}/\mathbb{P}^{q_w},$$

then we have

$$\text{span}(\mathcal{P}_{global}) = \mathbb{P}^{q_w+q_p} \oplus \xi^{q_p}\mathcal{W}_0 \oplus \ldots \oplus \xi\mathcal{W}_0 \oplus \mathcal{W}_0 = \mathcal{S}^{q_w+q_p, q_w-1}.$$

Manifold-Based B-Splines

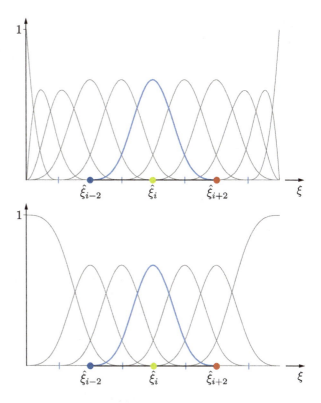

Fig. 5 Spline blending functions with $q_w = 3$ without boundary correction (top), with boundary correction (bottom)

The dimension of $\xi^k \mathcal{W}_0$ is independent of k and is equal to the number n of inner knots of the spline space $\mathcal{W} = \mathcal{S}^{q_w, q_w-1}$, i.e., $\dim(\mathcal{W}) = n + q_w + 1$, $\dim(\mathcal{W}_0) = n$ and

$$\dim(\text{span}(\mathcal{P}_{global})) = (q_w + q_p + 1) + n \cdot (q_p + 1).$$

As the span of $\mathcal{P}_{global}$ contains splines of degree $q_w + q_p$, we can expect an approximation order of $O(h^{q_w + q_p + 1})$ in L^2. Note that since the space of blending functions contains polynomials, the functions in $\mathcal{P}_{global}$ are linearly dependent for $q_w > 0$, as we then have $|\mathcal{P}_{global}| = (q_p + 1) \cdot (n + q_w + 1) > (q_w + q_p + 1) + n \cdot (q_p + 1)$.

The treatment of the boundary is not straightforward. In Fig. 5 two different options for choosing cubic blending functions with $q_w = 3$ is presented. The treatment of boundaries becomes relevant when extending the construction to surfaces. This leads to surfaces that are C^{q_w-1} if the mesh is regular. The construction on the top contains all splines, but leads to C^0 smooth surfaces at the extraordinary vertices. The construction on the bottom reproduces only a subspace of all B-splines, but generates surfaces that are C^2 smooth everywhere. However, as the overlap between the charts is large, the construction becomes cumbersome. Especially, in the bivariate case there can be several extraordinary vertices within one chart domain which can render their smooth parametrisation challenging.

For this reason, in the following, we consider blending functions that have a small support, but generate a smooth basis.

3.1.3 Rational B-Spline Blending Functions

To circumvent the difficulties that arise from using chart domains with a large number of overlaps, we construct blending functions that lead to only two overlapping chart domains in any point of the domain. That is, we consider blending functions with an overlap similar to the hat function in Fig. 4 (left), but possess the smoothness of higher order B-spline blending functions as in Fig. 5 (bottom). To reproduce this behaviour, we first select a linear combination of the B-splines in Fig. 5 as blending functions, so that the supports of no more than two blending functions overlap at the same time. In addition the B-splines to be used as blending functions are chosen from a suitably scaled coordinate system. See Fig. 6 for a construction with cubic B-splines with $q_w = 3$ defined on a coordinate axis scaled by a factor 2. Here, the blending functions are defined as

$$w_i(\xi) = \frac{B^3_{2i}(2\xi)}{B^3_{2(i-1)}(2\xi) + B^3_{2i}(2\xi) + B^3_{2(i+1)}(2\xi)}, \tag{16}$$

where B^3_k are the B-spline basis functions. Note that every other B-spline is chosen as a blending function and their sum does not add up to one. To obtain a partition of unity, all functions are divided by their sum. The resulting blending functions are then piecewise rational, as depicted in Fig. 6. The resulting manifold-based basis functions are also rational and their numerical integration may need more quadrature points than polynomial basis functions of similar order and smoothness.

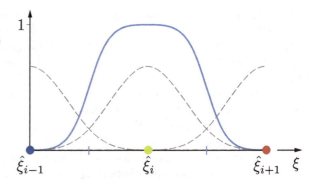

Fig. 6 Blending function w_i (solid line) as a normalised cubic B-spline (dashed lines)

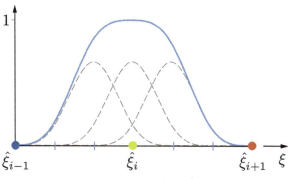

Fig. 7 Blending function w_i (solid line) composed out of three consecutive cubic B-splines (dashed lines) and the mask (1, 1, 1)

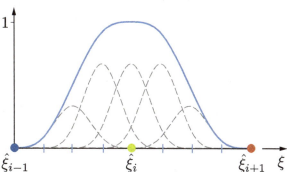

Fig. 8 Blending function w_i^* (solid line) composed out of five consecutive cubic B-splines (dashed lines) and the mask ($\frac{1}{2}$, 1, 1, 1, $\frac{1}{2}$)

3.1.4 Linear Combinations of B-Splines as Blending Functions

One can take linear combinations of consecutive B-splines as blending functions to obtain polynomial blending functions. We consider only cubic B-splines with $q_w = 3$ and express the coefficients in the linear combination as masks $(m_0, m_1, \ldots, m_k)$. To begin with, the B-splines to be used as blending functions are defined on a coordinate axis scaled by a factor 3 and the blending function mask is (1, 1, 1), see Fig. 7. That is, the blending functions are obtained as

$$w_i(\xi) = B_{3i}^3(3\xi) + B_{3i+1}^3(3\xi) + B_{3i+2}^3(3\xi). \tag{17}$$

Alternatively, it is possible to use B-splines defined on a coordinate axis scaled by a factor 4 with a mask ($\frac{1}{2}$, 1, 1, 1, $\frac{1}{2}$) so that

$$w_i^*(\xi) = \frac{1}{2}B_{4i}^3(4\xi) + B_{4i+1}^3(4\xi) + B_{4i+2}^3(4\xi) + B_{4i+3}^3(4\xi) + \frac{1}{2}B_{4i+4}^3(4\xi), \tag{18}$$

This choice is illustrated in Fig. 8. As is indicated in Figs. 7 and 8 within a chart the blending function $w(\xi)$ and $w^*(\xi)$ have five and seven breaking points, respectively.

Fig. 9 Global basis for a B-spline blending function composed according to (17) and a local cubic Bézier basis with $q_p = 3$ restricted to one chart

We can now compute the local contribution of the global basis on one chart. This is depicted in Fig. 9. The basis functions using weights as in (16) or (18) are visually indistinguishable and are omitted here.

3.1.5 Comparison of Different Blending Function Choices

To summarise, the constructions using rational functions or sums of B-splines generate charts that have only a one ring overlap. The same is true for piecewise linear hat functions as blending functions. Moreover, the dimension of the global function space is the same in these three cases. However, for piecewise linears, the functions are only C^0, whereas for the other three approaches the smoothness is C^{q_p-1}. In the following we compare the approximation power of the respective approaches by means of a numerical example.

In Fig. 10 we show log-error plots when performing L^2-fitting onto a given function. The corresponding number of degrees of freedom for each construction is given in Table 1. In this example we considered a sine function over the unit interval. The mesh size satisfies $h = 1/2^{\ell+2}$, where we used levels $\ell = 1, \ldots, 4$. We compare linear blending functions with local polynomials of degree $q_p = 2$ (blue line) and $q_p = 3$ (red line), as in Sect. 3.1.1. The former has a theoretical convergence rate of $O(h^4)$ in L^2, while the latter has a theoretical rate of $O(h^5)$. Both discretizations are C^0 only. We moreover compare rational blending functions as in (16) (yellow line), linear combinations of B-splines as blending functions as in (17) (green line) or in (18) (purple line) of degree $q_w = 3$ and local polynomials of degree $q_p = 3$. In all three cases, the expected convergence rate is $O(h^4)$.

We compare all constructions with uniform cubic B-splines of mesh size h (orange line). They can be interpreted as a manifold construction as in Sect. 3.1.2 with $q_w = 3$ and $q_p = 0$. This construction yields the highest error. Similarly high errors are observed for the basis from Sect. 3.1.1 with $q_p = 2$ (resulting in piecewise cubics) and the lowest error for the same construction with $q_p = 3$ (resulting in piecewise quartics). The bases constructed in Sect. 3.1.2 produce rates depending on the polynomial degree of the resulting splines. As all splines of a given degree and varying smoothness converge similarly (here e.g. $q_w = 2$, $q_p = 1$), we have omitted this case in Fig. 10. Note that the weight functions with boundary correction

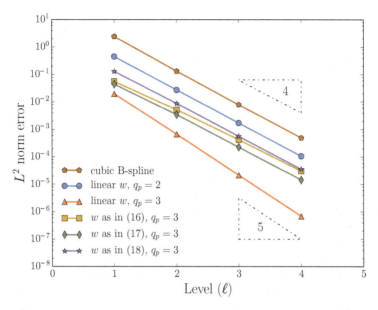

Fig. 10 L^2 approximation error plots for various choices of blending functions. Rates for h^4 and h^5 are included for comparison

Table 1 The number of degrees of freedom for the constructions compared in Fig. 10

	Number of DOFs
Cubic B-spline	$4 \cdot 2^\ell + 3$
Linear w, $q_p = 2$	$3(4 \cdot 2^\ell + 1)$
Linear w, $q_p = 3$	$4(4 \cdot 2^\ell + 1)$
w as in (16), $q_p = 3$	$4(4 \cdot 2^\ell + 1)$
w as in (17), $q_p = 3$	$4(4 \cdot 2^\ell + 1)$
w as in (18), $q_p = 3$	$4(4 \cdot 2^\ell + 1)$

in Fig. 5 (bottom) will not converge optimally without increasing the degree of local functions close to the boundary.

When comparing the constructions from Sects. 3.1.3 and 3.1.4, it turns out that all three converge with optimal rates of order h^4 and with significantly smaller constant when compared to piecewise polynomials of degree 3. This means that, when fitting onto a smooth, univariate function, manifold constructions yield a better approximation than standard B-splines, even though the manifold constructions do not reproduce B-splines. Among the manifold constructions, the non-rational variant (17) from Sect. 3.1.4 seems to be the faster.

Even though the numerical evidence suggests that manifold-based splines compare reasonably well with standard B-splines, there are still several open questions. On the one hand, one may compare the constructions with respect to computation times, which depend on the number of degrees of freedom as well as on the support size of the basis functions (i.e. the number of basis functions that are non-zero within one element). On the other hand, it is of vital importance to devise optimal quadrature rules, depending on the choice of blending functions. An efficient implementation for manifold-based B-splines, taking into account their specific structure, remains a task for future research.

3.2 Local Approximants

We discuss next how to choose the local basis $p_i^{(j)}$ so that the manifold-based construction reproduces B-splines of maximum smoothness. Here, the blending functions w_i have only to satisfy the partition of unity property. Hence, any one of the blending functions introduced in Sect. 3.1 can be used. To avoid the complications arising from chart domains $\hat{\Omega}_i$ with large number of overlaps, we consider only blending functions which lead to two overlapping charts. In addition, for the sake of concreteness we focus in the following on cubic B-splines and note that the proposed construction carries over to arbitrary degree.

The global manifold-based approximant (7) on a parameter domain consisting of a single finite interval is given by

$$f(\xi) = \sum_i w_i(\xi) f_i(\xi) = \sum_i w_i(\xi) \left(\sum_j p_i^{(j)}(\xi) \alpha_i^{(j)} \right). \tag{19}$$

It is required that this approximant is equal to a B-spline over all or some of the chart domains $\hat{\Omega}_i \equiv \operatorname{supp} \omega_i$. The cubic B-spline approximant is defined as

$$f^B(\xi) = \sum_k B_k^3(\xi) \beta_k, \tag{20}$$

where β_k are the control point coefficients. This approximant can be expressed as a weighted sum of chart domain contributions by multiplying with the partition of unity function, that is,

$$f^B(\xi) = \underbrace{\sum_i w_i(\xi)}_{\equiv 1} \sum_k B_k^3(\xi) \beta_k = \sum_i w_i(\xi) f_i^B(\xi) \tag{21}$$

The support of each of the terms $w_i(\xi) f_i^B(\xi)$ is strictly restricted to one chart domain, see Fig. 11. Note that the local basis in Fig. 11 (bottom) consists of five functions, which are scaled differently due to the multiplication with the blending function. Term by term matching of the manifold-based (19) and the weighted

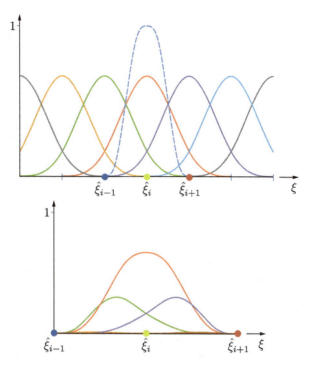

Fig. 11 Cubic B-spline basis functions and blending function (top). The weighted cubic B-spline basis functions on one chart to be reproduced with manifold-based basis functions (bottom)

B-spline approximants (21) requires on every chart domain

$$\sum_j p_i^{(j)} \alpha_i^{(j)} = f_i^B(\xi) \quad \text{for } \xi \in \hat{\Omega}_i = [\hat{\xi}_{i-1}, \hat{\xi}_{i+1}]. \tag{22}$$

This equation yields a set of equations for determining the coefficients $\alpha_i^{(j)}$ in dependence of the coefficients β_k. The cubic B-spline $B_k^3(\xi)$ has one knot with C^k, $k \leq 2$, at the centre of the chart domain $\xi = \hat{\xi}_i$ so that $p_i^{(j)}$ has to consist out of two pieces. Indeed, it is sufficient to consider in each half $[\hat{\xi}_{i-1}, \hat{\xi}_i]$ and $[\hat{\xi}_i, \hat{\xi}_{i+1}]$ of the chart domain $\hat{\Omega}_i$ a separate polynomial approximant. Choosing in each half a Bézier basis the coefficients α_i can simply be obtained by Bézier extraction as a linear combination of the B-spline coefficients β_k. Or more generally, the coefficients α_i are obtained by solving a small linear system of equations obtained by collocating (22) at four distinct points (for a cubic B-spline) within the segment. Since $f_i^B(\xi)$ is not known (22) has to be considered for each of the four non-zero B-spline basis functions individually. Similar to (9), this gives a relation between the two sets of coefficients expressed as

$$\boldsymbol{\alpha}_i = \mathbf{A}\mathbf{P}_i \boldsymbol{\beta}, \tag{23}$$

Note that the projection matrix **A** depends on the specific local basis chosen and is here the same on all the chart domains. Introducing the obtained coefficients into (19) yields the manifold-based basis functions, which are by design the same as the B-spline basis functions.

In the bivariate case the B-spline approximant (20) is only available on parts of the mesh with a tensor-product structure. In the vicinity of extraordinary vertices there is no representation as in (22). In such regions, as in the original manifold construction introduced in Sect. 2.2, a continuous polynomial approximant has to be fitted to the control mesh coefficients. The manifold construction ensures that the global approximant has the desired smoothness properties.

4 Conclusions

We developed new manifold-based B-spline basis functions motivated by the manifold-based surface construction techniques from geometric modelling. As illustrated the manifold-based surface construction techniques can be understood as the extension of the partition of unity method to manifolds. Specific choices for the blending functions and local approximants yield B-splines on structured control meshes. Owing to the flexibility of the partition of unity method several such choices are possible. We introduced in total five different choices for the blending functions two of which reproduce B-splines. In addition, we introduced one choice for the local approximant that leads to B-splines.

In Table 2 the properties of the manifold-based basis functions obtained from each of the six different choices are listed. For finite elements polynomial basis functions are to be preferred because they usually require fewer quadrature points to integrate. The number of breaking points within a finite element gives out of how many smoothly attached pieces a basis function consists. For efficient numerical integration the breaking points of the basis function have to be considered so that constructions with fewer breaking points are to be preferred. In the multivariate case, on unstructured meshes only constructions which require only one-ring of elements around each vertex as a chart domain are viable. If the chart domain

Table 2 Comparison of the properties of the introduced manifold-based basis functions on structured control meshes

	Rational/ polynomial	B-spline reproducing	Chart size	Breaking points	Approx. order
w in Sect. 3.1.1	p	✓	One-ring	0	h^{q_p+2}
w in Sect. 3.1.2	p	✓	$(q_w+1)/2$-ring	0	$h^{q_w+q_p+1}$
w in Sect. 3.1.3	r		One-ring	1	h^{q_p+1}
w in Sect. 3.1.4	p		One-ring	2	h^{q_p+1}
w^* in Sect. 3.1.4	p		One-ring	3	h^{q_p+1}
p as in Sect. 3.2	p	✓	One-ring	≥ 1	h^{q_p+1}

consists out of more than one ring of elements, there can be several extraordinary vertices in a chart which makes their parametrisation challenging. In the regular setting the approximation order of the introduced constructions can be inferred from Melenk and Babuska [15]. In Table 2 the higher order convergence of the first two constructions is remarkable. The first construction yields however only C^0 basis functions and the second construction requires charts with several rings of elements. Overall, the most promising constructions for finite elements appear to be the blending functions assembled from B-splines introduced in Sect. 3.1.4 and the local approximant introduced in Sect. 3.2. We note that the smoothness of the two resulting basis functions is C^k with $k = \min(q_p, q_w)$. In closing, we note that the mathematical and numerical study of the introduced constructions on unstructured meshes provides a promising area for future research.

Acknowledgments The authors would like to acknowledge the kind hospitality of the Erwin Schrödinger International Institute for Mathematics and Physics (ESI), where some of this research was carried out as part of the thematic programme *Numerical Analysis of Complex PDE Models in the Sciences*.

References

1. Babuška, I., Melenk, J.M.: The partition of unity method. International Journal for Numerical Methods in Engineering **40**, 727–758 (1997)
2. Buchegger, F., Jüttler, B., Mantzaflaris, A.: Adaptively refined multi-patch b-splines with enhanced smoothness. Applied Mathematics and Computation **272**, 159–172 (2016)
3. do Carmo, M.P.: Differential geometry of curves and surfaces. Prentice-Hall, Englewood Cliffs, NJ (1976)
4. Cirak, F., Long, Q.: Subdivision shells with exact boundary control and non-manifold geometry. International Journal for Numerical Methods in Engineering **88**, 897–923 (2011)
5. Cirak, F., Ortiz, M., Schröder, P.: Subdivision surfaces: A new paradigm for thin-shell finite-element analysis. International Journal for Numerical Methods in Engineering **47**, 2039–2072 (2000)
6. Collin, A., Sangalli, G., Takacs, T.: Analysis-suitable G^1 multi-patch parametrizations for C^1 isogeometric spaces. Computer Aided Geometric Design **47**, 93–113 (2016)
7. Cottrell, J.A., Hughes, T.J., Bazilevs, Y.: Isogeometric analysis: toward integration of CAD and FEA. John Wiley & Sons (2009)
8. Della Vecchia, G., Jüttler, B., Kim, M.S.: A construction of rational manifold surfaces of arbitrary topology and smoothness from triangular meshes. Computer Aided Geometric Design **25**, 801–815 (2008)
9. Grimm, C.M., Hughes, J.F.: Modeling surfaces of arbitrary topology using manifolds. In: SIGGRAPH 1995 Conference Proceedings, pp. 359–368 (1995)
10. Hughes, T.J.R., Cottrell, J.A., Bazilevs, Y.: Isogeometric analysis: CAD, finite elements, NURBS, exact geometry and mesh refinement. Computer Methods in Applied Mechanics and Engineering **194**, 4135–4195 (2005)
11. Kapl, M., Buchegger, F., Bercovier, M., Jüttler, B.: Isogeometric analysis with geometrically continuous functions on planar multi-patch geometries. Computer Methods in Applied Mechanics and Engineering **316**, 209–234 (2017)
12. Kapl, M., Sangalli, G., Takacs, T.: Analysis-suitable C^1 multi-patch isogeometric spaces: basis and dual basis. arXiv preprint arXiv:1711.05161 (2017)

13. Kapl, M., Sangalli, G., Takacs, T.: Construction of analysis-suitable G^1 planar multi-patch parameterizations. Computer-Aided Design **97**, 41–55 (2018)
14. Majeed, M., Cirak, F.: Isogeometric analysis using manifold-based smooth basis functions. Computer Methods in Applied Mechanics and Engineering **316**, 547–567 (2017)
15. Melenk, J.M., Babuska, I.: The partition of unity finite element method: Basic theory and applications. Computer Methods in Applied Mechanics and Engineering **139**, 289–314 (1996)
16. Navau, J.C., Garcia, N.P.: Modeling surfaces from meshes of arbitrary topology. Computer Aided Geometric Design **17**, 643–671 (2000)
17. Nguyen, T., Karčiauskas, K., Peters, J.: C^1 finite elements on non-tensor-product 2d and 3d manifolds. Applied Mathematics and Computation **272**, 148–158 (2016)
18. Peters, J., Reif, U.: Subdivision Surfaces. Springer Series in Geometry and Computing. Springer (2008)
19. Sangalli, G., Takacs, T., Vázquez, R.: Unstructured spline spaces for isogeometric analysis based on spline manifolds. Computer Aided Geometric Design **47**, 61–82 (2016)
20. Schutz, B.F.: Geometrical methods of mathematical physics. Cambridge University Press, Cambridge, UK (1980)
21. Scott, M.A., Simpson, R.N., Evans, J.A., Lipton, S., Bordas, S.P., Hughes, T.J., Sederberg, T.W.: Isogeometric boundary element analysis using unstructured T-splines. Computer Methods in Applied Mechanics and Engineering **254**, 197–221 (2013)
22. Scott, M.A., Thomas, D.C., Evans, E.J.: Isogeometric spline forests. Computer Methods in Applied Mechanics and Engineering **269**, 222–264 (2014)
23. Toshniwal, D., Speleers, H., Hiemstra, R.R., Hughes, T.J.: Multi-degree smooth polar splines: A framework for geometric modeling and isogeometric analysis. Computer Methods in Applied Mechanics and Engineering **316**, 1005–1061 (2017)
24. Toshniwal, D., Speleers, H., Hughes, T.J.: Smooth cubic spline spaces on unstructured quadrilateral meshes with particular emphasis on extraordinary points: Geometric design and isogeometric analysis considerations. Computer Methods in Applied Mechanics and Engineering **327**, 411–458 (2017)
25. Ying, L., Zorin, D.: A simple manifold-based construction of surfaces of arbitrary smoothness. In: SIGGRAPH 2004 Conference Proceedings, pp. 271–275 (2004)
26. Zhang, Q., Cirak, F.: Manifold-based isogeometric analysis basis functions with prescribed sharp features. Computer Methods in Applied Mechanics and Engineering **359**, 112659 (2020)
27. Zhang, Q., Sabin, M., Cirak, F.: Subdivision surfaces with isogeometric analysis adapted refinement weights. Computer-Aided Design **102**, 104–114 (2018)

Editorial Policy

1. Volumes in the following three categories will be published in LNCSE:

i) Research monographs
ii) Tutorials
iii) Conference proceedings

Those considering a book which might be suitable for the series are strongly advised to contact the publisher or the series editors at an early stage.

2. Categories i) and ii). Tutorials are lecture notes typically arising via summer schools or similar events, which are used to teach graduate students. These categories will be emphasized by Lecture Notes in Computational Science and Engineering. **Submissions by interdisciplinary teams of authors are encouraged.** The goal is to report new developments – quickly, informally, and in a way that will make them accessible to non-specialists. In the evaluation of submissions timeliness of the work is an important criterion. Texts should be well-rounded, well-written and reasonably self-contained. In most cases the work will contain results of others as well as those of the author(s). In each case the author(s) should provide sufficient motivation, examples, and applications. In this respect, Ph.D. theses will usually be deemed unsuitable for the Lecture Notes series. Proposals for volumes in these categories should be submitted either to one of the series editors or to Springer-Verlag, Heidelberg, and will be refereed. A provisional judgement on the acceptability of a project can be based on partial information about the work: a detailed outline describing the contents of each chapter, the estimated length, a bibliography, and one or two sample chapters – or a first draft. A final decision whether to accept will rest on an evaluation of the completed work which should include

- at least 100 pages of text;
- a table of contents;
- an informative introduction perhaps with some historical remarks which should be accessible to readers unfamiliar with the topic treated;
- a subject index.

3. Category iii). Conference proceedings will be considered for publication provided that they are both of exceptional interest and devoted to a single topic. One (or more) expert participants will act as the scientific editor(s) of the volume. They select the papers which are suitable for inclusion and have them individually refereed as for a journal. Papers not closely related to the central topic are to be excluded. Organizers should contact the Editor for CSE at Springer at the planning stage, see *Addresses* below.

In exceptional cases some other multi-author-volumes may be considered in this category.

4. Only works in English will be considered. For evaluation purposes, manuscripts may be submitted in print or electronic form, in the latter case, preferably as pdf- or zipped ps-files. Authors are requested to use the LaTeX style files available from Springer at http://www.springer.com/gp/authors-editors/book-authors-editors/manuscript-preparation/5636 (Click on LaTeX Template → monographs or contributed books).

For categories ii) and iii) we strongly recommend that all contributions in a volume be written in the same LaTeX version, preferably LaTeX2e. Electronic material can be included if appropriate. Please contact the publisher.

Careful preparation of the manuscripts will help keep production time short besides ensuring satisfactory appearance of the finished book in print and online.

5. The following terms and conditions hold. Categories i), ii) and iii):

Authors receive 50 free copies of their book. No royalty is paid.
Volume editors receive a total of 50 free copies of their volume to be shared with authors, but no royalties.

Authors and volume editors are entitled to a discount of 40 % on the price of Springer books purchased for their personal use, if ordering directly from Springer.

6. Springer secures the copyright for each volume.

Addresses:

Timothy J. Barth
NASA Ames Research Center
NAS Division
Moffett Field, CA 94035, USA
barth@nas.nasa.gov

Michael Griebel
Institut für Numerische Simulation
der Universität Bonn
Wegelerstr. 6
53115 Bonn, Germany
griebel@ins.uni-bonn.de

David E. Keyes
Mathematical and Computer Sciences
and Engineering
King Abdullah University of Science
and Technology
P.O. Box 55455
Jeddah 21534, Saudi Arabia
david.keyes@kaust.edu.sa

and

Department of Applied Physics
and Applied Mathematics
Columbia University
500 W. 120 th Street
New York, NY 10027, USA
kd2112@columbia.edu

Risto M. Nieminen
Department of Applied Physics
Aalto University School of Science
and Technology
00076 Aalto, Finland
risto.nieminen@aalto.fi

Dirk Roose
Department of Computer Science
Katholieke Universiteit Leuven
Celestijnenlaan 200A
3001 Leuven-Heverlee, Belgium
dirk.roose@cs.kuleuven.be

Tamar Schlick
Department of Chemistry
and Courant Institute
of Mathematical Sciences
New York University
251 Mercer Street
New York, NY 10012, USA
schlick@nyu.edu

Editor for Computational Science
and Engineering at Springer:

Martin Peters
Springer-Verlag
Mathematics Editorial IV
Tiergartenstrasse 17
69121 Heidelberg, Germany
martin.peters@springer.com

Lecture Notes
in Computational Science
and Engineering

1. D. Funaro, *Spectral Elements for Transport-Dominated Equations.*
2. H.P. Langtangen, *Computational Partial Differential Equations.* Numerical Methods and Diffpack Programming.
3. W. Hackbusch, G. Wittum (eds.), *Multigrid Methods V.*
4. P. Deuflhard, J. Hermans, B. Leimkuhler, A.E. Mark, S. Reich, R.D. Skeel (eds.), *Computational Molecular Dynamics: Challenges, Methods, Ideas.*
5. D. Kröner, M. Ohlberger, C. Rohde (eds.), *An Introduction to Recent Developments in Theory and Numerics for Conservation Laws.*
6. S. Turek, *Efficient Solvers for Incompressible Flow Problems.* An Algorithmic and Computational Approach.
7. R. von Schwerin, *Multi Body System SIMulation.* Numerical Methods, Algorithms, and Software.
8. H.-J. Bungartz, F. Durst, C. Zenger (eds.), *High Performance Scientific and Engineering Computing.*
9. T.J. Barth, H. Deconinck (eds.), *High-Order Methods for Computational Physics.*
10. H.P. Langtangen, A.M. Bruaset, E. Quak (eds.), *Advances in Software Tools for Scientific Computing.*
11. B. Cockburn, G.E. Karniadakis, C.-W. Shu (eds.), *Discontinuous Galerkin Methods.* Theory, Computation and Applications.
12. U. van Rienen, *Numerical Methods in Computational Electrodynamics.* Linear Systems in Practical Applications.
13. B. Engquist, L. Johnsson, M. Hammill, F. Short (eds.), *Simulation and Visualization on the Grid.*
14. E. Dick, K. Riemslagh, J. Vierendeels (eds.), *Multigrid Methods VI.*
15. A. Frommer, T. Lippert, B. Medeke, K. Schilling (eds.), *Numerical Challenges in Lattice Quantum Chromodynamics.*
16. J. Lang, *Adaptive Multilevel Solution of Nonlinear Parabolic PDE Systems.* Theory, Algorithm, and Applications.
17. B.I. Wohlmuth, *Discretization Methods and Iterative Solvers Based on Domain Decomposition.*
18. U. van Rienen, M. Günther, D. Hecht (eds.), *Scientific Computing in Electrical Engineering.*
19. I. Babuška, P.G. Ciarlet, T. Miyoshi (eds.), *Mathematical Modeling and Numerical Simulation in Continuum Mechanics.*
20. T.J. Barth, T. Chan, R. Haimes (eds.), *Multiscale and Multiresolution Methods.* Theory and Applications.
21. M. Breuer, F. Durst, C. Zenger (eds.), *High Performance Scientific and Engineering Computing.*
22. K. Urban, *Wavelets in Numerical Simulation.* Problem Adapted Construction and Applications.
23. L.F. Pavarino, A. Toselli (eds.), *Recent Developments in Domain Decomposition Methods.*

24. T. Schlick, H.H. Gan (eds.), *Computational Methods for Macromolecules: Challenges and Applications.*

25. T.J. Barth, H. Deconinck (eds.), *Error Estimation and Adaptive Discretization Methods in Computational Fluid Dynamics.*

26. M. Griebel, M.A. Schweitzer (eds.), *Meshfree Methods for Partial Differential Equations.*

27. S. Müller, *Adaptive Multiscale Schemes for Conservation Laws.*

28. C. Carstensen, S. Funken, W. Hackbusch, R.H.W. Hoppe, P. Monk (eds.), *Computational Electromagnetics.*

29. M.A. Schweitzer, *A Parallel Multilevel Partition of Unity Method for Elliptic Partial Differential Equations.*

30. T. Biegler, O. Ghattas, M. Heinkenschloss, B. van Bloemen Waanders (eds.), *Large-Scale PDE-Constrained Optimization.*

31. M. Ainsworth, P. Davies, D. Duncan, P. Martin, B. Rynne (eds.), *Topics in Computational Wave Propagation.* Direct and Inverse Problems.

32. H. Emmerich, B. Nestler, M. Schreckenberg (eds.), *Interface and Transport Dynamics.* Computational Modelling.

33. H.P. Langtangen, A. Tveito (eds.), *Advanced Topics in Computational Partial Differential Equations.* Numerical Methods and Diffpack Programming.

34. V. John, *Large Eddy Simulation of Turbulent Incompressible Flows.* Analytical and Numerical Results for a Class of LES Models.

35. E. Bänsch (ed.), *Challenges in Scientific Computing - CISC 2002.*

36. B.N. Khoromskij, G. Wittum, *Numerical Solution of Elliptic Differential Equations by Reduction to the Interface.*

37. A. Iske, *Multiresolution Methods in Scattered Data Modelling.*

38. S.-I. Niculescu, K. Gu (eds.), *Advances in Time-Delay Systems.*

39. S. Attinger, P. Koumoutsakos (eds.), *Multiscale Modelling and Simulation.*

40. R. Kornhuber, R. Hoppe, J. Périaux, O. Pironneau, O. Wildlund, J. Xu (eds.), *Domain Decomposition Methods in Science and Engineering.*

41. T. Plewa, T. Linde, V.G. Weirs (eds.), *Adaptive Mesh Refinement – Theory and Applications.*

42. A. Schmidt, K.G. Siebert, *Design of Adaptive Finite Element Software.* The Finite Element Toolbox ALBERTA.

43. M. Griebel, M.A. Schweitzer (eds.), *Meshfree Methods for Partial Differential Equations II.*

44. B. Engquist, P. Lötstedt, O. Runborg (eds.), *Multiscale Methods in Science and Engineering.*

45. P. Benner, V. Mehrmann, D.C. Sorensen (eds.), *Dimension Reduction of Large-Scale Systems.*

46. D. Kressner, *Numerical Methods for General and Structured Eigenvalue Problems.*

47. A. Boriçi, A. Frommer, B. Joó, A. Kennedy, B. Pendleton (eds.), *QCD and Numerical Analysis III.*

48. F. Graziani (ed.), *Computational Methods in Transport.*

49. B. Leimkuhler, C. Chipot, R. Elber, A. Laaksonen, A. Mark, T. Schlick, C. Schütte, R. Skeel (eds.), *New Algorithms for Macromolecular Simulation.*

50. M. Bücker, G. Corliss, P. Hovland, U. Naumann, B. Norris (eds.), *Automatic Differentiation: Applications, Theory, and Implementations.*
51. A.M. Bruaset, A. Tveito (eds.), *Numerical Solution of Partial Differential Equations on Parallel Computers.*
52. K.H. Hoffmann, A. Meyer (eds.), *Parallel Algorithms and Cluster Computing.*
53. H.-J. Bungartz, M. Schäfer (eds.), *Fluid-Structure Interaction.*
54. J. Behrens, *Adaptive Atmospheric Modeling.*
55. O. Widlund, D. Keyes (eds.), *Domain Decomposition Methods in Science and Engineering XVI.*
56. S. Kassinos, C. Langer, G. Iaccarino, P. Moin (eds.), *Complex Effects in Large Eddy Simulations.*
57. M. Griebel, M.A Schweitzer (eds.), *Meshfree Methods for Partial Differential Equations III.*
58. A.N. Gorban, B. Kégl, D.C. Wunsch, A. Zinovyev (eds.), *Principal Manifolds for Data Visualization and Dimension Reduction.*
59. H. Ammari (ed.), *Modeling and Computations in Electromagnetics: A Volume Dedicated to Jean-Claude Nédélec.*
60. U. Langer, M. Discacciati, D. Keyes, O. Widlund, W. Zulehner (eds.), *Domain Decomposition Methods in Science and Engineering XVII.*
61. T. Mathew, *Domain Decomposition Methods for the Numerical Solution of Partial Differential Equations.*
62. F. Graziani (ed.), *Computational Methods in Transport: Verification and Validation.*
63. M. Bebendorf, *Hierarchical Matrices. A Means to Efficiently Solve Elliptic Boundary Value Problems.*
64. C.H. Bischof, H.M. Bücker, P. Hovland, U. Naumann, J. Utke (eds.), *Advances in Automatic Differentiation.*
65. M. Griebel, M.A. Schweitzer (eds.), *Meshfree Methods for Partial Differential Equations IV.*
66. B. Engquist, P. Lötstedt, O. Runborg (eds.), *Multiscale Modeling and Simulation in Science.*
67. I.H. Tuncer, Ü. Gülcat, D.R. Emerson, K. Matsuno (eds.), *Parallel Computational Fluid Dynamics 2007.*
68. S. Yip, T. Diaz de la Rubia (eds.), *Scientific Modeling and Simulations.*
69. A. Hegarty, N. Kopteva, E. O'Riordan, M. Stynes (eds.), *BAIL 2008 – Boundary and Interior Layers.*
70. M. Bercovier, M.J. Gander, R. Kornhuber, O. Widlund (eds.), *Domain Decomposition Methods in Science and Engineering XVIII.*
71. B. Koren, C. Vuik (eds.), *Advanced Computational Methods in Science and Engineering.*
72. M. Peters (ed.), *Computational Fluid Dynamics for Sport Simulation.*
73. H.-J. Bungartz, M. Mehl, M. Schäfer (eds.), *Fluid Structure Interaction II - Modelling, Simulation, Optimization.*
74. D. Tromeur-Dervout, G. Brenner, D.R. Emerson, J. Erhel (eds.), *Parallel Computational Fluid Dynamics 2008.*
75. A.N. Gorban, D. Roose (eds.), *Coping with Complexity: Model Reduction and Data Analysis.*

76. J.S. Hesthaven, E.M. Rønquist (eds.), *Spectral and High Order Methods for Partial Differential Equations*.

77. M. Holtz, *Sparse Grid Quadrature in High Dimensions with Applications in Finance and Insurance*.

78. Y. Huang, R. Kornhuber, O.Widlund, J. Xu (eds.), *Domain Decomposition Methods in Science and Engineering XIX*.

79. M. Griebel, M.A. Schweitzer (eds.), *Meshfree Methods for Partial Differential Equations V*.

80. P.H. Lauritzen, C. Jablonowski, M.A. Taylor, R.D. Nair (eds.), *Numerical Techniques for Global Atmospheric Models*.

81. C. Clavero, J.L. Gracia, F.J. Lisbona (eds.), *BAIL 2010 – Boundary and Interior Layers, Computational and Asymptotic Methods*.

82. B. Engquist, O. Runborg, Y.R. Tsai (eds.), *Numerical Analysis and Multiscale Computations*.

83. I.G. Graham, T.Y. Hou, O. Lakkis, R. Scheichl (eds.), *Numerical Analysis of Multiscale Problems*.

84. A. Logg, K.-A. Mardal, G. Wells (eds.), *Automated Solution of Differential Equations by the Finite Element Method*.

85. J. Blowey, M. Jensen (eds.), *Frontiers in Numerical Analysis - Durham 2010*.

86. O. Kolditz, U.-J. Gorke, H. Shao, W. Wang (eds.), *Thermo-Hydro-Mechanical-Chemical Processes in Fractured Porous Media - Benchmarks and Examples*.

87. S. Forth, P. Hovland, E. Phipps, J. Utke, A. Walther (eds.), *Recent Advances in Algorithmic Differentiation*.

88. J. Garcke, M. Griebel (eds.), *Sparse Grids and Applications*.

89. M. Griebel, M.A. Schweitzer (eds.), *Meshfree Methods for Partial Differential Equations VI*.

90. C. Pechstein, *Finite and Boundary Element Tearing and Interconnecting Solvers for Multiscale Problems*.

91. R. Bank, M. Holst, O. Widlund, J. Xu (eds.), *Domain Decomposition Methods in Science and Engineering XX*.

92. H. Bijl, D. Lucor, S. Mishra, C. Schwab (eds.), *Uncertainty Quantification in Computational Fluid Dynamics*.

93. M. Bader, H.-J. Bungartz, T. Weinzierl (eds.), *Advanced Computing*.

94. M. Ehrhardt, T. Koprucki (eds.), *Advanced Mathematical Models and Numerical Techniques for Multi-Band Effective Mass Approximations*.

95. M. Azaïez, H. El Fekih, J.S. Hesthaven (eds.), *Spectral and High Order Methods for Partial Differential Equations ICOSAHOM 2012*.

96. F. Graziani, M.P. Desjarlais, R. Redmer, S.B. Trickey (eds.), *Frontiers and Challenges in Warm Dense Matter*.

97. J. Garcke, D. Pflüger (eds.), *Sparse Grids and Applications – Munich 2012*.

98. J. Erhel, M. Gander, L. Halpern, G. Pichot, T. Sassi, O. Widlund (eds.), *Domain Decomposition Methods in Science and Engineering XXI*.

99. R. Abgrall, H. Beaugendre, P.M. Congedo, C. Dobrzynski, V. Perrier, M. Ricchiuto (eds.), *High Order Nonlinear Numerical Methods for Evolutionary PDEs - HONOM 2013*.

100. M. Griebel, M.A. Schweitzer (eds.), *Meshfree Methods for Partial Differential Equations VII*.

101. R. Hoppe (ed.), *Optimization with PDE Constraints - OPTPDE 2014*.

102. S. Dahlke, W. Dahmen, M. Griebel, W. Hackbusch, K. Ritter, R. Schneider, C. Schwab, H. Yserentant (eds.), *Extraction of Quantifiable Information from Complex Systems*.

103. A. Abdulle, S. Deparis, D. Kressner, F. Nobile, M. Picasso (eds.), *Numerical Mathematics and Advanced Applications - ENUMATH 2013*.

104. T. Dickopf, M.J. Gander, L. Halpern, R. Krause, L.F. Pavarino (eds.), *Domain Decomposition Methods in Science and Engineering XXII*.

105. M. Mehl, M. Bischoff, M. Schäfer (eds.), *Recent Trends in Computational Engineering - CE2014*. Optimization, Uncertainty, Parallel Algorithms, Coupled and Complex Problems.

106. R.M. Kirby, M. Berzins, J.S. Hesthaven (eds.), *Spectral and High Order Methods for Partial Differential Equations - ICOSAHOM'14*.

107. B. Jüttler, B. Simeon (eds.), *Isogeometric Analysis and Applications 2014*.

108. P. Knobloch (ed.), *Boundary and Interior Layers, Computational and Asymptotic Methods – BAIL 2014*.

109. J. Garcke, D. Pflüger (eds.), *Sparse Grids and Applications – Stuttgart 2014*.

110. H. P. Langtangen, *Finite Difference Computing with Exponential Decay Models*.

111. A. Tveito, G.T. Lines, *Computing Characterizations of Drugs for Ion Channels and Receptors Using Markov Models*.

112. B. Karazösen, M. Manguoğlu, M. Tezer-Sezgin, S. Göktepe, Ö. Uğur (eds.), *Numerical Mathematics and Advanced Applications - ENUMATH 2015*.

113. H.-J. Bungartz, P. Neumann, W.E. Nagel (eds.), *Software for Exascale Computing - SPPEXA 2013-2015*.

114. G.R. Barrenechea, F. Brezzi, A. Cangiani, E.H. Georgoulis (eds.), *Building Bridges: Connections and Challenges in Modern Approaches to Numerical Partial Differential Equations*.

115. M. Griebel, M.A. Schweitzer (eds.), *Meshfree Methods for Partial Differential Equations VIII*.

116. C.-O. Lee, X.-C. Cai, D.E. Keyes, H.H. Kim, A. Klawonn, E.-J. Park, O.B. Widlund (eds.), *Domain Decomposition Methods in Science and Engineering XXIII*.

117. T. Sakurai, S.-L. Zhang, T. Imamura, Y. Yamamoto, Y. Kuramashi, T. Hoshi (eds.), *Eigenvalue Problems: Algorithms, Software and Applications in Petascale Computing*. EPASA 2015, Tsukuba, Japan, September 2015.

118. T. Richter (ed.), *Fluid-structure Interactions*. Models, Analysis and Finite Elements.

119. M.L. Bittencourt, N.A. Dumont, J.S. Hesthaven (eds.), *Spectral and High Order Methods for Partial Differential Equations ICOSAHOM 2016*. Selected Papers from the ICOSAHOM Conference, June 27-July 1, 2016, Rio de Janeiro, Brazil.

120. Z. Huang, M. Stynes, Z. Zhang (eds.), *Boundary and Interior Layers, Computational and Asymptotic Methods BAIL 2016*.

121. S.P.A. Bordas, E.N. Burman, M.G. Larson, M.A. Olshanskii (eds.), *Geometrically Unfitted Finite Element Methods and Applications*. Proceedings of the UCL Workshop 2016.

122. A. Gerisch, R. Penta, J. Lang (eds.), *Multiscale Models in Mechano and Tumor Biology*. Modeling, Homogenization, and Applications.

123. J. Garcke, D. Pflüger, C.G. Webster, G. Zhang (eds.), *Sparse Grids and Applications - Miami 2016*.

124. M. Schäfer, M. Behr, M. Mehl, B. Wohlmuth (eds.), *Recent Advances in Computational Engineering*. Proceedings of the 4th International Conference on Computational Engineering (ICCE 2017) in Darmstadt.

125. P.E. Bjørstad, S.C. Brenner, L. Halpern, R. Kornhuber, H.H. Kim, T. Rahman, O.B. Widlund (eds.), *Domain Decomposition Methods in Science and Engineering XXIV*. 24th International Conference on Domain Decomposition Methods, Svalbard, Norway, February 6–10, 2017.

126. F.A. Radu, K. Kumar, I. Berre, J.M. Nordbotten, I.S. Pop (eds.), *Numerical Mathematics and Advanced Applications – ENUMATH 2017*.

127. X. Roca, A. Loseille (eds.), *27th International Meshing Roundtable*.

128. Th. Apel, U. Langer, A. Meyer, O. Steinbach (eds.), *Advanced Finite Element Methods with Applications*. Selected Papers from the 30th Chemnitz Finite Element Symposium 2017.

129. M. Griebel, M.A. Schweitzer (eds.), *Meshfree Methods for Partial Differencial Equations IX*.

130. S. Weißer, BEM-based Finite Element *Approaches on Polytopal Meshes*.

131. V.A. Garanzha, L. Kamenski, H. Si (eds.), *Numerical Geometry, Grid Generation and Scientific Computing*. Proceedings of the 9th International Conference, NUMGRID2018/Voronoi 150, Celebrating the 150th Anniversary of G. F. Voronoi, Moscow, Russia, December 2018.

132. H. van Brummelen, A. Corsini, S. Perotto, G. Rozza (eds.), *Numerical Methods for Flows*.

133. H. van Brummelen, C. Vuik, M. Möller, C. Verhoosel, B. Simeon, B. Jüttler (eds.), *Isogeometric Analysis and Applications 2018*.

134. S.J. Sherwin, D. Moxey, J. Peiro, P.E. Vincent, C. Schwab (eds.), *Spectral and High Order Methods for Partial Differential Equations ICOSAHOM 2018*.

135. G.R. Barrenechea, J. Mackenzie (eds.), *Boundary and Interior Layers, Computational and Asymptotic Methods BAIL 2018*.

136. H.-J. Bungartz, S. Reiz, B. Uekermann, P. Neumann, W.E. Nagel (eds.), *Software for Exascale Computing - SPPEXA 2016–2019*.

137. M. D'Elia, M. Gunzburger, G. Rozza (eds.), *Quantification of Uncertainty: Improving Efficiency and Technology*.

138. ——

139. F.J. Vermolen, C. Vuik (eds.), *Numerical Mathematics and Advanced Applications ENUMATH 2019*.

For further information on these books please have a look at our mathematics catalogue at the following URL: www.springer.com/series/3527

Monographs in Computational Science and Engineering

1. J. Sundnes, G.T. Lines, X. Cai, B.F. Nielsen, K.-A. Mardal, A. Tveito, *Computing the Electrical Activity in the Heart.*

For further information on this book, please have a look at our mathematics catalogue at the following URL: www.springer.com/series/7417

Texts in Computational Science and Engineering

1. H. P. Langtangen, *Computational Partial Differential Equations.* Numerical Methods and Diffpack Programming. 2nd Edition
2. A. Quarteroni, F. Saleri, P. Gervasio, *Scientific Computing with MATLAB and Octave.* 4th Edition
3. H. P. Langtangen, *Python Scripting for Computational Science.* 3rd Edition
4. H. Gardner, G. Manduchi, *Design Patterns for e-Science.*
5. M. Griebel, S. Knapek, G. Zumbusch, *Numerical Simulation in Molecular Dynamics.*
6. H. P. Langtangen, *A Primer on Scientific Programming with Python.* 5th Edition
7. A. Tveito, H. P. Langtangen, B. F. Nielsen, X. Cai, *Elements of Scientific Computing.*
8. B. Gustafsson, *Fundamentals of Scientific Computing.*
9. M. Bader, *Space-Filling Curves.*
10. M. Larson, F. Bengzon, *The Finite Element Method: Theory, Implementation and Applications.*
11. W. Gander, M. Gander, F. Kwok, *Scientific Computing: An Introduction using Maple and MATLAB.*
12. P. Deuflhard, S. Röblitz, *A Guide to Numerical Modelling in Systems Biology.*
13. M. H. Holmes, *Introduction to Scientific Computing and Data Analysis.*
14. S. Linge, H. P. Langtangen, *Programming for Computations* - A Gentle Introduction to Numerical Simulations with MATLAB/Octave.
15. S. Linge, H. P. Langtangen, *Programming for Computations* - A Gentle Introduction to Numerical Simulations with Python.
16. H.P. Langtangen, S. Linge, *Finite Difference Computing with PDEs* - A Modern Software Approach.
17. B. Gustafsson, *Scientific Computing from a Historical Perspective.*
18. J. A. Trangenstein, *Scientific Computing.* Volume I - Linear and Nonlinear Equations.

19. J. A. Trangenstein, *Scientific Computing*. Volume II - Eigenvalues and Optimization.
20. J. A. Trangenstein, *Scientific Computing*. Volume III - Approximation and Integration.

For further information on these books please have a look at our mathematics catalogue at the following URL: www.springer.com/series/5151

CPSIA information can be obtained
at www.ICGtesting.com
Printed in the USA
LVHW082047180121
676820LV00001B/3